Globalization and Environmental Reform

Globalization and Environmental Reform

The Ecological Modernization of the Global Economy

Arthur P. J. Mol

The MIT Press
Cambridge, Massachusetts
London, England

Set in Sabon by The MIT Press.
Printed and bound in the United States of America.

Library of Congress Cataloging-in-Publication Data

Mol, A. P. J.
Globalization and environmental reform : the ecological modernization of the global economy / Arthur P. J. Mol.
p. cm.
Includes bibliographical references and index.
ISBN 0-262-13395-4 (hc. : alk. paper)
1. Sustainable development. 2. Environmental policy. 3. Environmental degradation. 4. Globalization. I. Title.
HC79.E5 .M64 2001

2001030440

to Jan and Marijke

Contents

Preface

When notions of globalization appear in environmental debates and publications, it is not often in conjunction with phrases such as "environmental reform" and "ecological modernization." Environmental issues are more often portrayed as the dark side of globalization, with the environment as one of the victims of ongoing neo-liberal globalization. Increasing liberalization of trade and investment, decreasing control by nation-states of what is happening within their territories, the growing power of transnational companies—all these factors contribute to a rather apocalyptic view of the environment in globalization debates.

This volume investigates the other side of globalization. Are there settings in which globalization enhances environmental reform and contributes to the ecological modernization of the global economy? If so, what dynamics, actors, and institutions are responsible, and what can we learn from them about designing and implementing more sustainable systems of global production and consumption? Are these opportunities distributed equally around the world. What are the social drawbacks for peripheral regions of environmental reform of the global economy?

Portions of the book have been presented at conferences and workshops or have been published as articles and book chapters. The book has benefited from the comments of those who have heard or read them. The following conferences proved to be excellent discussion platforms for raising some of the ideas put forward herein: Environmental Risks and the Quality of life (Niterói, Brazil, 1996), Social Theory and the Environment (Woudschoten, the Netherlands, 1997), World Congress of Sociology, RC24 Session on Ecological Modernization (Montreal, 1998), The Environmental State under Pressure (Chicago, 1999), and Globalization, Governance and the Environment (Berkeley, 2000). I am most grateful to the organizers and participants of these gatherings for fruitful discussions, critical reflections, and new insights.

I am especially indebted to several colleagues who have been co-authors of various publications that have laid the foundation for this book. Gert Spaargaren has not only been my most important partner in reflection; he also commented on all chapters and greatly helped me to improve my line of argument. David Sonnenfeld was invaluable for "testing" ecological modernization ideas outside Europe. His intellectual support, Asian expertise, and substantial comments on earlier drafts of several chapters have substantially improved the book. Fred Buttel has sharpened my ideas on ecological modernization with his comments and reflections. Bas van Vliet and Jos Frijns have been extremely helpful with their research and insights into Vietnam, Kenya, and the Netherlands Antilles.

Special thanks also go to Kris van Koppen, Rob Kuitenbrouwer, Duncan Liefferink, Lloyd Narain, Phung Thuy Phuong, Corry Rothuizen, and Zhang Lei. Clay Morgan at The MIT Press greatly helped me in turning the manuscript into a book.

Globalization and Environmental Reform

1

Beyond Seattle

Late in November of 1999, a wide variety of pressure groups and concerned citizens gathered in Seattle to stage a protest during a ministerial meeting of the World Trade Organization (WTO) at which 135 countries aimed to set the agenda for a new round of trade liberalization. It proved to be a memorable demonstration, and not only for its violent character. During previous negotiations on world trade (for example, the last Uruguay round), the debate on trade liberalization had been the monopoly of professional negotiators. At Seattle, an unprecedented number of demonstrators participated in the debate. The coalition of non-governmental organizations (NGOs) opposing the WTO's new trade-liberalization proposals was also unprecedented. More than 1500 NGOs had signed an anti-WTO declaration in advance. More than 770 NGOs were actually present, among them environmental NGOs, labor unions, Third World and fair-trade NGOs, extreme leftist groups, consumer organizations, farmer organizations, and human rights activists. Also new was the common target: globalization.

'Globalization' appears to have become a catchword, much as 'capitalism' was in the late 1960s and the 1970s. Food safety, environmental protection, biodiversity, protection of children against misuse as laborers, human rights, democracy, and opposition to capitalism all seem only aspects of one common purpose: rejecting globalization.

Not only did the Seattle ministerial meeting fail to formulate a common agenda for future negotiations on trade liberalization; most commentators also agreed that the WTO, the epitome of globalization, would never be the same after Seattle. The agreement on the Biosafety Protocol (which covers food products produced with genetically modified organisms, and which was negotiated within the framework of the Biodiversity Convention), concluded only 2 months later, strongly reflects the protests against what was then labeled globalization. This protocol was received favorably by environmental organizations, by the European Union's Commissioner for the Environment, Margot Wallström, and by a large number of developing countries. It allows countries to limit imports of food products based on genetically modified organisms, and it asks that such products be labeled. Above all, the protocol clearly states that it cannot be overruled by the WTO (although it remains to be seen if this will hold in practice). But does the Biosafety Protocol mean less globalization? Alternatively, is it actually globalization, but of a different (regulated) kind? Or is it just a small ripple in a continuing stream of globalization that is devastating our environment? These and related questions are the central themes of this book, which aims to analyze and understand why globalization provoked such widely differing reactions during the "Battle of Seattle." Globalization was the common object of vehement criticism by the protesters; at the same time, it was applauded for its economic, environmental, and social benefits by most of the governments of OECD (Organization for Economic Cooperation and Development) countries, by the WTO itself, and by the chief executive officers of transnational companies.

I aim to shed light on the often confusing and contradictory arguments put forward by proponents and opponents (both academics and politicians) of globalization, particularly in regard to the assumed environmental gains and hazards. As an environmental sociologist, I sympathize with the environmental NGOs in regard to the environmental impacts of globalization. Nevertheless, my main intention here is to look beyond the idea of a pervasive process of globalization that destroys the environment. My intent is not so much to contradict the protesters at Seattle and subsequent summits, who were correct to point out the potentially disastrous effects of the trade liberalization advocated by some WTO members. Rather, I want to show that it is not globalization that these pressure groups attack and reject, and that it should not be. Globalization, I will argue and illustrate, is a multifaceted phenomenon with potentially devastating but also potentially beneficial environmental consequences. The environmental NGOs were particularly afraid—and not without reason—of globalization in its one-sided, neo-liberal, economic aspect.

Understanding Anti-Globalization Perspectives

The word 'globalization' first became fashionable in the late 1980s. In the mid 1980s it was not even in the vocabulary of the same politicians who now speak of a need to break down rules and regulations in order to make national industries more competitive in a globalizing world, or of those who argue in favor of national protection to prevent the breakdown of a nation's culture, industry, environmental quality, or economy by global forces. It is equally difficult to imagine those who speak for large multinational corporations and major international economic organizations not referring to globalization when demanding liberalization, privatization, and the lifting of "protective" measures. In the mid 1980s, a wide variety of environmental NGOs, labor movements, Third World groups, and fair-trade groups supported the notion of "sustainable development," but they had not identified a common "enemy." In Seattle and at numerous other summits, they joined forces under the banner of anti-globalization. All this is evidence of the rapid dissemination of the idea of globalization.

Globalization: What Is New about It?

'Globalization' may be a rather new word, but its predecessors 'internationalization' and 'transnationalization' were widely used from the nineteenth century to the late 1980s. But is 'globalization' just a more fashionable word or concept, or does its emergence reflect a new condition in worldwide economic, political, and/or cultural relations? Though most people may use 'globalization' unconsciously, not comparing it with other terms, academics should be more deliberate in selecting or avoiding it. Although all concepts point at processes and activities that extend beyond national boundaries, there are two clear differences between globalization and the other two:

• The traditional concept of internationalization is considered too narrow for the current times, as it focuses too much on the increasingly interwoven nature of *national* economies and *nation-states* through international trade and political relations. In the 1990s it became generally accepted that nation-states are not the only—and according to some, not even the most important—actors in global processes. Multinational companies of various kinds, financial organizations, global networks of NGOs, sub-national regions, and cities play important roles in shaping worldwide developments.

Similarly, the conventional notion of transnationalization was limited to the development of organizations and production on a cross-border basis. It focused particularly on multinational organizations such as those of the United Nations. These two "old" concepts continue to depict countries or nation-states as the building blocks of all global processes—an idea that is seen as increasingly inadequate by those who prefer the notion of globalization. WTO meetings, such as the one in Seattle, are no longer of interest only to representatives of nation-states; they are increasingly of concern to representatives of businesses and of environmental and other NGOs.

• Globalization is better suited to the social developments of ever-intensifying and ever-extending networks of cross-border human interaction. Such interaction bridges increasing distances in decreasing time, distinguishing the present era of globalization from the era of internationalization. Examples of this accelerating time-space compression are not found only in the financial markets (connected as they now are by satellites and the Internet, and circulating immense amounts of financial capital around the globe instantly, beyond the control of any private or public agency). Local environmental NGOs also benefit from global interconnectedness. In 1999 we witnessed their ability to share information rapidly by means of the World Wide Web when an effective global campaign against the WTO ministerial meeting in Seattle was organized in only 2 months. In 1998 they fought successfully against the Multilateral Agreement on Investment as drafted by the OECD, and in early 2000 they protested the gathering of globalization proponents at the annual meeting of the World Economic Forum in Davos, Switzerland.

The notion of globalization has advanced forcefully on political, public, and research agendas, having acquiring a firm position in the early 1990s.[1] Consciously or not, many academic scholars, political commentators, politicians, business representatives, and NGOs prefer the concept of globalization to the conventional ideas of internationalization and transnationalization. Globalization reflects a kind of common-sense view of the global transformations and the interdependence that most people claim (or are told) to witness. Increasing numbers of people have discarded the idea of the nation-state as the rule, the organizing principle, and the unit, and of everything outside it as an exception. This is particularly important for environmentalists, environmental problems, and environmental reform in OECD countries, where for at least 25 years environmentalists have relied heavily (and, more than incidentally, in vain) on strong nation-states. Environmentalists have always been hostile to infringements on the nation-state, whether they be due to privatization, to deregulation, or, now, to globalization.

With the widespread and rapid introduction of the notion of globalization came vehement criticism of both the concept and the processes of globalization. In what follows, I will focus on two interrelated sets of anti-globalization arguments that, together, provide a better understanding of this criticism and balance the anti-globalization debate.

Anti-Globalization 1: The Myth of Globalization

The farthest-reaching conclusion drawn by some academics, politicians, and non-governmental representatives from the confusing interpretations and unclear definitions of globalization is the denial that such a phenomenon as globalization exists. Defenders of this position are found not so much within the ranks of NGOs that gathered in Seattle to protest against the WTO, or within the ranks of non-governmental visitors to the numerous meetings of the Conference of the Parties aiming to influence the United Nations Framework Convention on Climate Change (UNFCCC). These groups are all too aware that a new global framework is in the making. The main arguments against the globalization thesis are perhaps best summarized by the social scientists Paul Hirst and Grahame Thompson (1995), the most outspoken and the clearest of those who deny that anything new is going on.[2] Referring to the "myth of globalization," Hirst and Thompson summarize their main ideas as follows (ibid., pp. 195–196):

We have pointed out the following problems for the globalisation thesis: first, that few exponents of globalisation develop a coherent concept of the world economy in which supranational forces and agents are decisive; second, that pointing to evidence of the enhanced internationalisation of economic relationships since the 1970s is not in itself proof of the emergence of a distinctly 'global' economic structure; third, that the international economy has been subject to major structural changes in the last century and that there have been earlier periods of internationalisation of trade, capital flows and the monetary system, especially in 1870–1914; fourth, that truly global TNCs are relatively few and that most successful multinational corporations continue to operate from distinct national bases; and lastly, that the prospects for regulation by international cooperation, the formation of trading blocks, and the development of new national strategies that take account of internationalisation are by no means exhausted.

Such views are valuable in that they correct some important aspects of the confusion that has emerged in the globalization debate and in that they counterbalance the idea that a completely new phenomenon has emerged out of the blue. Globalization certainly does not mean the end of the nation-state as a powerful actor in the international arena. National governments, individually or collectively, still exercise sovereign authority on various issues, not least on a considerable number of environmental problems. However, denying that anything new is happening is going one step too far, as I will show throughout this book. Those who speak of the "myth of globalization" often have a clear political strategy underlying their analyses and their ideas. Their main purpose is to attack the unfettered global market economy, with its widespread negative social consequences and external effects, and to make a strong case for the still considerable opportunities for nation-states to regulate and counteract the forces of global capitalism.

Those who deny that such a thing as globalization exists often do so to defend the nation-state against powerful economic forces. One might sympathize with this political perspective for several reasons:

• The nation-state system is one of the cornerstones of democratic governance. As such, it is a much more desirable normative basis for decision making on our common future than a global capitalist market.

• Nation-states have firmly established correcting mechanisms to counteract and neutralize the undesirable side effects of economic systems, including environmental problems.

• Some members of the political and business elites of advanced Western nations—staunch advocates of neo-liberal ideas and programs of privatization and deregulation—have found in globalization a convenient cloak for the domestic policies they have chosen to pursue. In the strategies of these elites, globalization is often glorified as the only way out of domestic troubles or presented as an inescapable fact of life that necessitates the adoption of adequate neo-liberal domestic policies if the national economy is to keep up in the "rat race" and retain its competitiveness with respect to other countries.

"Political" use of the concept of globalization and corresponding attacks on the nation-state have triggered severe criticism of globalization, especially by those opposed to neo-liberal political and economic programs. Indeed, Hirst and Thompson's analysis should be partly understood as exactly this: criticism of the ideologies that parallel political and economic globalization. And in this Hirst and Thompson are not alone. Amin (1997, p. 127) takes a position similar to that of many of the NGOs that protested against the WTO in Seattle. Their point is that big business, international regulatory organizations, and influential conservative governments believe and/or ideologically promote the idea that, under the present conditions of globalization, the only way to advance national and global prosperity is to apply the familiar cocktail of neo-liberal policies, including deregulation, privatization, reduced public expenditures, wage competition, unrestrained trade, market integration, and tight international control of exchange and interest rates. It is this neo-liberal project that they oppose. But rather than criticize globalization for its non-existence or irrelevance (as Hirst and Thompson seem to do), these critics protest against it or attack it because of the detrimental political and other consequences it entails.

Anti-Globalization 2: Global Capitalism

In spite of the differing positions taken in the globalization controversy, some shared issues have emerged. The relationship between capitalism and globalization is among the most prominent of these. Globalization is often closely associated with, if not limited to, the dynamics of global capitalism. To a large extent, the emergence, the shape, and the dynamics of what is called globalization are explained by the internal dynamics of the capitalist mode of production. Consequently, globalization will lead to the same kinds of social disasters that befell capitalism. Indeed, much of the severe criticism of globalization is related to this close connection. On a national scale, the industrialized societies have, to varying extents, managed to reduce and neutralize the most severe consequences of the free capitalist market. We are now witnessing the return of the very same problems on a global scale as we fail to achieve any successful correction at that level. It is globalization—or global capitalism—that is the root cause of a new round of destruction. Traditionally, neo-Marxists were convinced that, owing to the internal contradiction of capital and the capitalist mode of production, capitalism would eventually lead to its own destruction and dissolution.[3] Today most globalization scholars who focus on the economic dynamics of globalization have departed from such deterministic positions and acknowledge the flexibility of capitalism in overcoming these internal contradictions. For instance, Ankie Hoogvelt (1996, p. 132) concludes:

. . . capitalism, instead of destroying itself in consequence of internal contradictions which are its inherent and systemic properties, time and again proves itself able to overcome the self-inflicted crisis by complete transformation. Total renewal is what makes possible the reproduction of capitalism, involving not only production technology and the organization of economic life but also the complex of institutions and norms which ensure that individual agents and social groups behave according to the overarching principles of economic life.

One possible conclusion about the environmental consequences of global capitalism is that globalization may triumph over the internal economic contradictions of capitalism, but that it will lead to the end of the global capitalist economic order because it jeopardizes the sustenance base of the treadmill of production and consumption. This line of thinking, which was implicit in the ideas of some of the Seattle protesters, is articulated by neo-Marxist environmental sociologists such as Ted Benton, Peter Dickens, Allan Schnaiberg, and James O'Connor. They have combined the idea of aggressive global expansion of the capitalist economy with the ongoing and intensifying global environmental crisis to formulate the "second contradiction of capitalism" argument, according to which the economic growth and the economic expansion that are inherent in the global capitalist economy will run up against environmental boundaries that will, in the end, turn the tide of the global capitalist economic order and change it beyond recognition. Eric Hobsbawm reaches a similar conclusion on the environmental crisis in the last chapter of *Age of Extremes* (1994).

However, there is confusion among neo-Marxists with respect to the environmental consequences of global capitalism and the repercussions of the environmental crisis on global capitalism. O'Connor (1998, p. 235), a leading American neo-Marxist, concludes that "a systematic answer to the question, 'Is an ecologically sustainable capitalism possible?' is, 'Not unless and until capital changes its face in ways that would make it unrecognizable to bankers, money managers, venture capitalists, and CEOs looking at themselves in the mirror today.'" Dickens (1998, p. 191), a renowned European neo-Marxist, has a more balanced assessment: "According to this second contradiction argument, nature will continue to wreak 'revenge' on society as a result of capitalism. Several related questions remain, however. First, will capitalism be able to restructure itself once more, this time in the form of, what has been called, 'ecological modernization'?" Enrique Leff (1995) takes this one step further. From a neo-Marxist perspective, he initially rejects attempts to simply incorporate environmental concerns into global capitalist development through internalization of externalities and other strategies of ecological modernization. However, Leff finally reaches the conclusion that an environmentally sound development is not "totally incompatible with capitalist production" (ibid., p. 126).

The ambivalence of neo-Marxists may be the scientific equivalent of the ambivalence of environmental NGOs. Similarly, the environmental ideas of neo-liberal economists have exponents among the representatives of businesses, international economic organizations, and economic and industrial ministries of most OECD countries who gathered at the 2000 annual meeting of the World Economic Forum.

Balancing Anti-Globalization Perspectives

NGOs and their academic spokesmen criticize globalization primarily because they equate it with neo-liberalism. However, it can be questioned to what extent this equation is correct, insofar as it reduces the idea of globalization to a simple economic neo-liberalism operating beyond national borders. Globalization can refer to other social processes, including increased exchange of information by environmental NGOs around the world, enhanced global environmental politics, and global diffusion of environmental norms and values. Jan Nederveen Pieterse (1996, p. 5) is perhaps most to the point in asserting that this "critique of globalization . . . only targets a caricature ideology of globalization without addressing and in fact obfuscating the main issues" and in asking "If the target is neo-liberalism and the unfettered market-economy, why attack globalization?" If neo-liberal policies are sometimes politically and ideologically justified in appealing to globalization, it is going a step too far to summarily reject globalization as a concept for analyzing the kinds of changes that are now taking place in society.

That the relationship between globalization with neo-liberalism is more than a matter of semantics becomes clear when one takes its consequences into account. A tight connection between globalization and neo-liberalism reduces the range of possible solutions to contemporary problems related to neo-liberalism. It more or less forces the opponents of neo-liberal policies to refer back to some of the strategies traditionally used to counter neo-liberalism: strengthening sovereign nation-states, increasing domestic control of economic trade and investment, preventing any takeover of public tasks by the private sector. Concurrently, it limits the scope for new alternatives that better suit the contemporary settings of increasing global interdependence.

Starting from a position that acknowledges the global transformations now taking place but still rejects neo-liberal solutions to contemporary problems, one can come up with different political and economic answers to today's problems and challenges. Designing democratic global governance and forming a global civil society are processes that aim to control unrestrained global competition and neo-liberalism, not by denying or rejecting globalization, but by addressing the question of how to shape and design globalization and globalization processes in a desirable way. In such a perspective, environmental NGOs should not object to or reject globalization; they should attack certain elements or forms of a globalizing world order while strengthening others. This is the view that I will argue.

Sociological Perspectives on Globalization and the Environment: A Short Review of the Literature

A logical first step in analyzing how globalization interferes with the environment is to look at the existing literature. What does social science have to offer toward clarifying the relationship between globalization and the environment? How do social scientists relate globalization to environmental deterioration, environmental consciousness, and environmental reform? What elements of globalization could work in favor of environmental quality? My first general overview of the literature will draw on two bodies of sociological[4] literature: globalization studies and environmental sociology.

In an indirect way, the sociological literature on globalization brings forward relevant insights into the environmental question and discourse. In a review of several publications on globalization, Nederveen Pieterse (1997, p. 371) concludes that, although globalization theories come in a wide variety, most authors seem to agree on three points:

• Globalization is concentrated in the interlinked economies of Europe, North America, and Japan and Southeast Asia. (I will label these "the triad.")
• The "North-South" gap is narrowing for some high-performance economies (especially in Southeast and East Asia and in Latin America) but is widening for others (the "real" developing countries, particularly those in sub-Saharan Africa).
• Delinking has lost much of its appeal as a means of escaping the devastating consequences of economic globalization, particularly for developing countries or regions.[5]

The third conclusion adds to the observation that counterproductivity theories, which were dominant during the second wave of environmental concern, are becoming less prominent in the environmental debate. Evidently, the declining attractiveness of counterproductivity ideas is not only a result of an "internal" development within the environmental discourse or a follow-on from the emergence of environment-induced reform in institutions and practices at the level of the nation-state; it is also deeply interconnected with globalization trends and debates. This conclusion is also shared by Paul Wapner (1995), who discusses sub-statist and supra-statist responses to state failure in environmental reform. Wolfgang Sachs (1993) and his colleagues show that moderate delinking strategies still have advocates.

The two main conclusions to be drawn from the more direct references to the environment in the sociological literature on globalization are the following:

• Although direct references to the environment are regularly made in the globalization literature (Archer 1991; Mlinar 1992; Barber 1995; Held 1995; Dalby 1996; Spybey 1996; Nederveen Pieterse 1997), they are usually no more than statements. In most cases, a more detailed and profound analysis of the relationship between globalization in the world economy, in international politics, and in global culture (on one hand) and environmental deterioration, environmental perceptions, and environmental reform (on the other hand) is lacking. For instance, only in the completely revised third edition of *Global Shift,* a standard work on economic globalization, has Dicken added a small last chapter on the environmental governance of global economic trade and investment (Dicken 1998, pp. 463–467).[6] We are left to wonder about the reasons for these often superficial analyses. Are these linkages taken for granted and considered to be evident with no need of further explanation? Are they seen as relatively unimportant in understanding globalization? Alternatively, is there some practical or intellectual reason why globalization studies do not include more refined environmental analyses?

• In the few studies of globalization that deal more extensively with the environment, tendencies toward globalization are identified as a new causal factor of further environmental deterioration and environmental anxiety. These studies (e.g., Scholte 1996) hardly concentrate and elaborate on the potential benefits of globalization processes for, or its contribution to, global environmental management, governance, and reform.[7] Most merely mention the internationalization of the environmental movement.[8] There are a few influential exceptions, most notably Giddens (1990, 1991, 1998), Held et al. (1999), and Beck (1991, 1994, 1997); however, Held and Beck express rather apocalyptic views on the environmental consequences of globalization.

A review of the literature on environmental sociology reveals a slightly better but still disappointing record. Besides the observation that global environmental change rather than globalization receives the most attention, the general conclusions are as follows:

• Sociological theories of globalization are only marginally employed to investigate and understand the causes behind global environmental deterioration and change and to identify new forms of global environmental management and reform. Most writings on this topic in environmental sociology concentrate on analyzing concrete empirical phenomena rather than on developing a generally applicable theoretical understanding. There are many case studies of the environmental consequences of global economic phenomena (e.g., the WTO), of the emergence of environmentalism as a global culture (as manifested by international NGOs and international opinion polls), and of international environmental relations, law, and decision making related to environmental agreements and to organizations such as the United Nations and the European Union. Redclift and Benton (1994), Yearly (1996), Redclift (1996), and Wilenius (1999)[9] have published more theoretically inspired writings, but they are exceptions.

• Most contributions to the globalization debate in the fields of environmental sociology and the environmental sciences (e.g., Anton 1995) emphasize the negative and destructive consequences of globalization for the natural environment. This can be interpreted as an academic reflection of environmental NGOs, insofar as it seems to see no solution other than to resist globalization (Mittelman 1998). Again, globalization, often defined as global capitalism, is seen as a menace to the maintenance of global environmental quality.

The following two conclusions can be drawn from this initial overview of studies in the field of globalization and the environment:

• There are only few studies that provide an overall view of how globalization is related to the environment.

• The majority of these contributions analyze the relationship between globalization and the environment in rather negative terms, highlighting the environmental threats of globalization.

Outline of the Book

Most studies of globalization and the environment proceed from a strong normative point of view and argue either "that the impact of globalization is utterly destructive" or "that it is the royal road forward, that it represents the way of the future, or that it does not exist" (Nederveen Pieterse 1997, p. 371). Only recently, and only slightly, has the number of analytical studies of globalization that do not fall victim to one of these narrow positions increased. A large body of writing suggests (here I paraphrase Waterman 1996) that globalization and further economic development constitute a broad and winding highway to environmental hell, and that small and local alternatives form the straight and narrow path to ecological paradise. In contrast, there are works that seem to glorify globalization not only from an economic standpoint[10] but also from a standpoint of civil society and the environment. Such globalization studies do to all global alternatives what the World Commission on Environment and Development (the Brundtland Commission) has done to the ecological alternatives: they attempt to reconcile the dominant path of economic development with what have been identified as negative side effects of that same development path, and by so doing they try to incorporate alternative ideologies and scenarios into mainstream thinking. It is exactly this incorporation that is so vehemently disputed, not only by the neo-Marxists quoted above, but also by scholars such as Sachs (1993).

In this volume I attempt to transcend the good/bad dichotomy that has dominated the writings on globalization and the environment (and which, as I have noted, is still prominent in the literature). The purpose of the book is to provide a better and more balanced understanding of the relationship between globalization and environmental quality. This is especially relevant at the dawn of a century and a millennium, as Sachs et al. (1998, p. 8) conclude: "Come what may, the twenty-first century will be the century of the environment—either the century of ecological catastrophes or the century of ecological transformation."

Instead of glorifying or condemning globalization from the outset, I start from a more analytical perspective of globalization. Subsequently I analyze the interaction between the processes of globalization in its various forms and dimensions (on the one hand) and environmental degradation, perception, and reform (on the other). For two reasons, however, I emphasize the dialectics of globalization processes and environmental *reform* rather than environmental *degradation*. First, I want to balance the rather apocalyptic overtones that prevail in most contemporary studies on the environment and globalization. Second, I want to investigate the theory of ecological modernization of production and consumption (which until now has focused on the national level of European countries) and to determine its value in regard to understanding international and global environmental reforms.

The more theoretical part of the volume begins, in chapter 2, with a detailed review and interpretation of various theories of globalization. Here I take a discontinuist standpoint with respect to the globalization controversy, and I relate globalization to what is often labeled late, reflexive, or even global modernity (Spaargaren et al. 2000).

Chapter 3 starts at the other end, with the changing interpretations of environmental disturbances and environmental reforms. The theory of the ecological modernization of production and consumption, in particular, has been developed to conceptualize and understand the changes in the environmental debate as well as the changes in actual environmental practices. However, the ideas of most contributors to the ecological modernization theory have been confined to Western Europe until now. Only limited attention has been paid to global economic and political changes that have affected environmental protection and decay. In that sense, chapter 3 can be interpreted as an extended research question, albeit in more theoretical terms.

Chapter 4 addresses the interpretations of globalization that concentrate on its contributions to further environmental decay, to the decreasing possibilities of environmental management and control, and to an increasing sense of "apocalypse blindness" on the part of the forces that speed up global interconnections and dynamics. In short, it focuses on the dark side of globalization. In chapter 5, I try to conceptualize and elaborate on how globalization processes and dynamics can or do contribute (often unexpectedly or indirectly, sometimes intentionally) to environmental reforms and environmental consciousness. These two opposing perspectives are expounded in the three subsequent chapters, with an emphasis on environmental reform.

Chapter 6 is the beginning of the more empirical and less theoretical part of the book, which concerns the dialectics of globalization and environmental reform in various economic regions. Here I concentrate on the triad (Europe, North America, and Japan and East and Southeast Asia) and on the global institutions that are closely connected to these major regions of economic globalization. I clarify how globalization processes interact with environmental deterioration and reform in each of these regions, and on what innovative steps are being made in each region to turn neo-liberal economic globalization toward more environmentally sound directions.

Chapter 7 focuses on the environmental consequences for the less developed countries of environmental and economic reform in the economic triad. The World Trade Organization, the Multilateral Agreement on Investment, the various Multilateral Environmental Agreements, and the question of relocating pollution from areas with stringent environmental regimes to areas with lax controls will be analyzed.

Chapter 8 concentrates on three countries that exemplify different state-economy combinations, offering Vietnam as an example of a newly industrializing economy in Asia with specific transitional characteristics and a strong state, the Netherlands Antilles as an example of a "small island development state" with an open economy, and Kenya as an example of a predatory African state with an economy that is hardly integrated into the global economy.

In the final chapter, I reflect on the general and theoretical insights gained by this study. First, I summarize what an ecological modernization perspective has to offer us in regard to how globalization and environmental reforms interact. I then analyze the consequences of the globalization processes and dynamics for what is still a very Western or even European theory of ecological modernization.

2

Globalization and the Transformation of Modernity

From 1995 on, a considerable number of developed industrialized countries, working together under the banner of the Organization for Economic Cooperation and Development (OECD), began negotiations for a Multilateral Agreement on Investment (MAI). This investment agreement was supposed to become the equivalent of the existing trade agreement under the World Trade Organization. Designed to create a level playing field for global investment and to liberalize such investment, it was the OECD member states' political answer to the movement toward globalization in investments and increasing global interdependence. Representatives of the industrialized member states negotiated in isolation for 2 years before drafts were presented to those not involved in the negotiations. Many of those "excluded"—among them many NGOs and most of the developing nations—were not particularly happy with these drafts. The OECD and its member states argued that countries were free to choose whether to join the MAI once it had been accepted by the OECD member states. However, developing countries pointed to the fact that the terms of the agreement had been set by the most developed nations according to their priorities and requirements, and that the poorer countries had no choice but to put up with it. Environmentalists chiefly criticized the neo-liberal character of the treaty. They were concerned with the fact that nation-states would be deprived of a number of measures for protecting their countries against unwanted investments that were beneficial to multinational investors but not to the environment. Environmentalists argued that the agreement put nation-states at a disadvantage relative to multinational corporations. The rapidly intensifying debate that followed the release of the drafts of the MAI in 1997 resulted, in the autumn of 1998, in temporary rejection of the proposal, support for which was rapidly diminishing in various countries (although not among the political and economic elites in the triad).

Many of the critics of the 1998 draft of the MAI did not oppose such a treaty in principle; they opposed the specific contents and the closed process of negotiation. Most critics agreed that there was a movement toward globalization in investments and therefore a need to regulate these global investments (even if only to prevent a "law of the jungle" capitalism in which only the economically most powerful would survive, with all the related costs). The MAI was, however, believed to be too neo-liberal for such "regulation." In criticizing the MAI, one must ask whether the environmental regulation of economic processes and institutions can be dealt with (as happened during the twentieth century) by strengthening the nation-states or the collective system of nation-states. Perhaps the new conditions of the global economy require new systems and "regulations." Perhaps we should accept that individual nation-states will become less important regulators (although not necessarily for the benefit of transnational corporations) and that supra-national or even global political structures and arrangements need to be created and strengthened. To even begin to answer such questions, it is essential to have a more detailed idea of the new path along which the world is moving. Are the developments in the global economy marked by changes in trade, capital flows, and investments fundamentally different from those of 50 years ago? Do political and cultural institutions show similar global transformations?[1]

Globalization: "Continuist" and "Discontinuist" Interpretations

Although the notion of globalization has advanced rapidly among the sociological concepts designed to analyze the changing character of the modern world, there is still some debate on three somewhat interdependent points:

• To what extent is globalization a trend that can be identified in the real world?
• Is this process of globalization something new, or is it merely a continuation of a development that began some 400 or more years ago?
• What are the most important features of globalization?

In regard to the first question, Scholte (1996) identifies three main schools of thought: conservatives, who deny that a trend such as globalization exists; liberals (both neo-liberals and reformists), who celebrate what they regard as the fruits of globalization; and critics (including historical materialists, postmodernists, and poststructuralists), who decry its alleged disempowering effects.[2] The first group is part of the "realist" tradition of international relations, which insists on the central role of the state and the continuation of sovereignty in social development worldwide. Asserting that the contemporary debate on globalization is basically between reformists and its critics, Scholte (ibid., pp. 54–55) emphasizes the diminishing significance of the conservative standpoint on globalization. Held (1995, p. 25) joins Scholte, concluding that "there is not much evidence to suggest that realism and neo-realism possess a convincing account of the enmeshment of states within the wider global order, of the effects of the global order of states, and of the political implications of all this for the modern democratic state." If for the moment we define globalization loosely as the increasing stretching of social relationships beyond the level of the nation-state, most contributors to this debate would agree that it does not make much sense to deny the existence of those trends in the real world.

To an extent, the controversy as to whether globalization is a new phenomenon or just a continuation and an extension of processes of interdependence and interconnectedness that began with the emergence of capitalism some 400 years ago has replaced the discussion on the existence of globalization. Though most scholars now acknowledge that global interconnectedness and interdependence exist, and that it is worthwhile to analyze international or transnational economic, political, and societal processes in this light, they disagree on the continuity or discontinuity of these processes. One of the strongest defenders of continuity is Immanuel Wallerstein, who insists that the internationalization of the world economy is a continuous process that parallels the development of capitalism. Within the political sciences, Peter Gourevitch (1978) and others emphasize the constancy of local and international interdependence and interpenetration from the sixteenth century on. Starting from a predominantly political and cultural point of view, Roland Robertson supports the above-mentioned authors, identifying the fifteenth century as the starting point of globalization and the 1880s as the time when globalization accelerated. According to Robertson, the emergence of the idea of the homogeneous nation-state, the sharp increase in the number of international agencies and institutions, the increasing global forms of communication (railways, telephone, post), the acceptance of unified global time, the development of global competition, and the emergence of a standard notion of citizenship all contributed to the validity of this time frame. Gordon (1988), Glyn and Sutcliffe (1992), and Hirst and Thompson (1992) agree that we are not living in a new globalized economy, since the present situation is not fundamentally different from that of, say, 100 years ago. Glyn and Sutcliffe (1992, p. 91) assert that "the widespread view that the present degree of globalization is in some way new and unprecedented is, therefore, false." These authors acknowledge that globalization is a real trend but maintain that its core features date back as far as either the sixteenth or the nineteenth century. On this basis, they interpret 'globalization' as a popular contemporary catchword denoting something that has existed and has been studied for quite some time.

If I disagree with these so-called continuist analyses, I am not denying elements of continuity, and I acknowledge that most social developments do not come in the shape of sudden radical changes.[3] However, it should be emphasized that the forms and dynamics of interconnectedness and interdependence, which are so central to most notions of globalization, have changed fundamentally in the last 30–40 years.[4] Philip McMichael (1996b) is a typical advocate of the "discontinuist" school of thought in globalization studies. From a sociology-of-development perspective, McMichael convincingly restructures the development debate by making a strong distinction between what he calls the development project (which runs up to the 1970s) and the globalization project (which begins to become discernible in the 1980s). A second advocate of discontinuity, Anthony Giddens (1990), emphasizes the significance of the new telecommunication and information technologies, which allow acceleration in the compression of time and space and thus contribute to a qualitative change in globalization. These technologies and the related organizational innovations have altered the scope and the speed of economic decision making, thereby enhancing the capacity of the economic system to respond rapidly to fluctuations but also rendering it more vulnerable to overreaction to relatively minor disturbances. Moreover, the new communication and information technologies have not only affected the economic system; they have also influenced the political and cultural dimensions of modernity. Manuel Castells's trilogy *The Information Age* (1996, 1997a,b) contains a similar analysis, with a heavy emphasis on information flows and on information and communication technologies as signs of the coming of age of globalization. Looking at globalization from an economic point of view, Peter Dicken (1998) argues that the recently emerging globalization processes are qualitatively different from the older economic internationalization processes. Dicken holds that this has resulted in a new geo-economy in which national boundaries and nation-states are less and less relevant to the dynamics of world economics. Finally, in one of the first handbooks on globalization, Held, McGrew, Goldblatt, and Perraton (1999, pp. 430–431) combine several of the arguments mentioned above and provide a detailed list of eleven features that together distinguish between what they call the present phase of globalization from earlier phases of internationalization. The common position of all these scholars on the continuum between continuity and discontinuity has everything to do with the core features of this new era of globalization and with its relationship to the transformation of modernity.

"Old" and "New" Theories of Globalization

In a compact but remarkably complete and informative overview of the various sociological contributions to the issue of globalization, Malcolm Waters (1995) makes a useful distinction between "old" and "new" theories of globalization. This distinction is also helpful in analyzing the core features of globalization.

Waters summarizes the "old" theories of the third quarter of the twentieth century under four headings: modernization and convergence theories (centering on Parsonian functionalism and on post-industrialists such as Clark Kerr and Daniel Bell), world capitalism theories (especially those revolving around Wallerstein's World-System concept, Sklair 1991 being a more recent example),[5] international relations theories (especially those that depart from the realist framework and focus on interdependence, the theories of Morse (1976) and Keohane and Nye (1977) being among the first), and studies centering on Marshall McLuhan's notion of the global village. With the exception of some branches of World-System theories, these quite different schools of thought look as though they might converge on the underlying idea that something fundamental was changing in the modern world. The four branches of the "old" theories were developed quite separately, without much exchange, interaction, or cross-fertilization. These theories differ from their successors on three main counts:

• As McGrew (1992) and Yearly (1996) have shown, the search for one prime cause or trigger for the processes of globalization is what distinguishes these "old" theories from the "new" theories, which emphasize the multiplicity of causes.[6]
• The "old" theories were often not sufficiently able to identify a common denominator in the changes occurring in all institutional clusters of modern society.
• The "old" theories did not see these changes as a fundamental transformation or discontinuity in the historical development of capitalism, industrialism, the nation-state system, or any other institutional feature of modernity.

Moreover (and this is of special relevance for the present study), these theories were largely blind to environmental issues. The changes in the modern world system seemed to be independent of real or perceived environmental problems. Only recently have some contributors to each of these schools of thought tried to redress that omission.[7]

As Waters rightly states, Roland Robertson should be seen as the grandfather of the "new" globalization studies that made a major breakthrough in the late 1980s and the early 1990s. Robertson's papers, which began to appear in the mid 1980s, focused explicitly on the notion of globalization as an "umbrella" concept for the social processes and transformations that were covered separately by the "old" theories. Moving beyond monolithic theories of the international system of nation-states and beyond economic World-System theories, Robertson aimed to give a broader account of what is at stake in the global arena, emphasizing the relative independence of the cultural dimension. His classical definition of globalization points to the concrete processes of increasing interdependence and interconnectedness emerging around the world, and to the growing awareness of this: "Globalization as a concept refers both to the compression of the world and the intensification of consciousness of the world as a whole." (Robertson 1992, p. 8) Anthony Giddens is often seen as the second leading theorist on globalization, and his "definition" of globalization may be quoted even more often than Robertson's: "The intensification of worldwide social relations which link distant localities in such a way that local happenings are shaped by events occurring many miles away and vice versa." (Giddens 1990, p. 64) Though McGrew (1992) and others identify

Giddens as the chief representative of the recent upsurge in globalization studies and include Robertson in his school of thought, this is at least chronologically incorrect; according to some, it is also substantially incorrect.[8] This point of disagreement aside, it cannot be denied that Giddens has made a significant contribution to globalization theories. However, more than many other globalization theorists, Giddens linked globalization processes to the environment: "Ecological problems highlight the new and accelerating interdependence of global systems and bring home to everyone the depth of the connections between personal activity and planetary problems." (1991, p. 221)

There is one additional characteristic shared by many "new" theories on globalization. A return to the discourse on modernity can be witnessed not only in the work of Robertson and Giddens but also in most of the recent theoretical contributions to the globalization debate. This discourse has its roots in the debate on Parsonian modernization theory of the 1960s and the 1970s and in the post-modernity discourse of the early 1980s. Giddens, in his contributions to understanding globalization, basically expands on his earlier work on the transformation of modern society. Thus, if the "new" contributions share a preoccupation with modernity, they also engage in considerable debate and controversy in regard to the exact relationship between globalization and modernity. I will briefly elaborate on this relationship between globalization and modernity, as it seems essential for an understanding of the core features of globalization as well as for any investigation into its environmental dimensions.

The Transformation of Modernity

Homogenization versus Heterogenization

Analyzing the core features and transformative "powers" of globalization processes, various scholars have focused on the question of whether globalization processes have universalizing effects and will therefore result in homogenization (or even Westernization) of the world. On this view, globalization would create a world in which distinct societies would become more and more alike as a result of economic forces (e.g., global chains, networks, and flows in manufacturing, trade, and finance) and developments in the cultural domain (e.g., cultural imperialism, homogeneous cosmopolitans), in the political domain (diminishing capacity of the nation-state, world polity),[9] and in other domains. The standard would be the capitalist post-industrial society of the West, with its typical cultural and political outlook. Much in line with these homogenization ideas, Cohen (1996) supplies an example of the new convergence thesis in his analysis of the growing similarities, including environmental threats and reforms, between major cities in the North and in the South. Owing to various globalization processes, all major cities increasingly face similar environmental problems and threats and are thus in need of similar environmental reforms. If we are to believe these analysts, Beijing, Mexico City, London, and Los Angeles all face air pollution, traffic jams, environmental risks related to food products, and growing use of energy and drinking water and are all in need of efficient public transport systems, of effective environmental monitoring, control, and enforcement programs, and of renewable resources and demand-side management of natural resources. Robertson (1992) and Nederveen Pieterse (1995) have criticized this interpretation of globalization as a renewed form of the traditional Parsonian modernization theory. They claim that the concept of evolutionary convergence is theoretically inadequate, as has been discussed extensively in the debate on Parson's functionalism (Giddens 1984, pp. 263–274). In addition, the concept usually neglects the inherent dark side of modernity so colorfully portrayed by Zygmunt Bauman. Indeed, Bauman (1993, pp. 186–222) explicitly includes global environmental threats and failing environmental reforms in the dark side of an ambivalent modernity.

On the other side of the globalization debate, some have identified globalization as inherently leading to tendencies of heterogenization, or to diverging developments. Heterogenization can be identified in the economic field as unequal distribution of goods, money, and power; the number of studies illustrating or arguing this are countless. However, heterogenization is also related to the increasing inequality of access to environmental resources, to the unequal distribution of what Hans Opschoor (1992) has labeled "environmental space," and to the unequal abilities of societies or groups to deal with the environmental risks and threats they face. Globalization processes affect all economies on the planet, albeit in very dissimilar ways: either by inclusion in or by exclusion from the processes of production, circulation, and consumption, and either by high quality of life or by low quality of life. And economic exclusion often goes together with low quality of life.

Increasingly, these two opposite perspectives converge on the idea that, although there are strong global economic, political, and cultural forces at work in a comparable way all over the world, these forces have different local effects. McMichael (1996a, pp. 40–41) puts it as follows in his analysis of institutions that are transformed under the "power" of globalization: "Globalization is ultimately an institutional transformation. It has no single face, as institutions and institutional change vary across the world." We can witness these diverse consequences and outlooks in cultural, economic, and political institutions. One and the same global cultural message is received and interpreted differently in various localities. Environmentalists are familiar with the fact that the global cultural norm of sustainability has quite different meanings and interpretations in different contexts around the world. The same global cultural producer (e.g., CNN) often adapts its product to regional or local markets instead of sending one universal message. Similarly, global economic producers tailor their products and even their production systems to local and regional markets, preferences, and requirements. The economist Alan Rugman (1994) acknowledges that multinational enterprises will have to deal with the twin goals of globalization and national responsiveness to consumer tastes and government regulations. This means that their global activities must, to some extent, accommodate local conditions, both in terms of content and in terms of process. Others (e.g., Dicken 1997) emphasize the diversity of transnational companies, whose specific past and present circumstances in their home countries still strongly influence their organizational characteristics as well as their strategies around the world. Finally, global politics and policies do not have equal consequences for different countries and areas. National political circumstances and conditions "mediate" global political regimes. Even in more homogeneous regions that have truly supranational regimes, such as the European Union, diverging national and local political institutions force homogeneous tendencies to become more heterogeneous. Giddens (1994b, p. 188) summarizes this as follows: "Globalization can no longer be understood as Westernization; developing societies and developed societies mix in culture, economy and politics." Globalization can result in homogenization, in heterogenization, or in hybridization (Nederveen Pieterse 1995), depending on the specific configuration of actors and institutions in the relevant social system or domain. It is too simple to see globalization as a uniform process of Westernization or

McDonaldization resulting in an increasingly homogeneous single global system, or as a process causing increasingly diverging and heterogenizing effects in different parts of the world. In spite of this general consensus, there are differences in emphasis regarding the homogeneity and the heterogeneity of globalization.

Reflexive Modernization

The debate on homogeneous versus heterogeneous transformations due to globalization processes can be interpreted as a part of a wider controversy regarding the relationship between globalization and modernity. It is a controversy that involves Robertson and Giddens especially, although Nederveen Pieterse (1995) and others have contributed to it. Robertson (1995) criticizes Giddens (1990) for seeing globalization as a consequence of modernity, whereas it should be identified rather as a general condition that has facilitated modernity. This difference depends to a major extent on how globalization and modernity are defined and interpreted. For instance, Robertson seems to have a rather broad conception of globalization, seeing it as a process that began in the fifteenth or the sixteenth century, whereas Giddens restricts the notion of globalization to the last 40 years of the twentieth century. On the other hand, Robertson seems to restrict the notion of modernization to a rather Parsonian one[10] and is therefore able to criticize Giddens for both his narrow emphasis on modernity and his analysis of globalization as a consequence of modernity in a particular era. However, this seems a rather narrow interpretation and reading of Giddens's work on modernity and globalization. Giddens's notion of modernity is more analytical, and it departs from the normative and substantive connotations of Parsonian sociology that have been criticized so strongly by Giddens himself (1984, pp. 263–274) and by others. Giddens links the process of globalization to a specific phase of late-modern society by analyzing the transformations that have taken place in four institutional clusters: capitalism, industrialism, surveillance through the nation-state, and the military order.[11] Central or even prior to globalization processes in these four institutional clusters is the "liberation" of time and space, which is closely linked with disembedding: it is a process whereby social relations are lifted out of local contexts of interaction and recombined across time and space. Giddens, in his analysis of the emergence of globalization since the 1960s, points particularly to the contribution of modern communication and information technologies (these being part of the institutional cluster of industrialism) in shrinking the world. In particular, it is the growing speed of flows of information (and related flows of money, capital, culture, images, beliefs, ideas, and so on)[12] in the increasingly global networks of production and exchange that has brought about a qualitative change in the world in the past 40 years.

It is through the latter interpretation that globalization processes are linked to the analysis of the changing character of modernity and the emergence of a new phase in modernity, known as late or reflexive modernization. As Kumar (1995) has argued convincingly, globalization studies are among the last in a series of sociological contributions to define and interpret the changing constitution of modernity. These began with the post-industrial and information society theories of the 1970s and continued with the post-Fordist theories, the post-modernist, and the diverse reflexive modernization theories. Various themes raised in the discourse on globalization originated in these earlier theoretical contributions, including the emergence of new communication and information technologies, the transformations in the organization of production and consumption cycles, and the dialectics of global and local.[13] The connection between globalization and reflexive modernization is particularly strong. Radicalized or reflexive modernity is analyzed by Giddens (1990, 1991, 1994a), among others, as a phase marked by globalization and the end of tradition. 'Reflexivity' refers to the constant reexamining and reshaping of social practices in the light of new incoming information about those very practices; it thus denotes the end of the idea that social and natural environments will be increasingly subjected to rational ordering. The globalizing processes occurring during the era of reflexive modernity contribute to feelings of anxiety and uncertainty among citizens. At the institutional level, reflexivity involves the routine incorporation of new information and knowledge into social conduct and institutional arrangements, transforming the institutional order. Science and technology, the bureaucratic decision-making institutions at the political and administrative levels, and the organization of production-consumption cycles—all especially relevant for environmental deterioration and reforms—are transformed into modes of operation that differ from their predecessors in the era of simple or high modernity. Globalization plays a significant role in these transformation processes. Neither science and technology (or expert systems) nor national political arrangements can remain unchallenged when confronted by globalization processes. A similar line of reasoning is put forward by Scott Lash and Ulrich Beck. Beck makes the most direct connection between globalization and the changing character of modernity (on one hand) and the perception of ecological risks and related anxieties (on the other hand). The persistent environmental threats and the chronic environmental anxieties experienced by large segments of society can no longer be counteracted by the institutional mechanisms of simple or high modernity. Globalization processes are among the main causes of this disparity. Castells (1997a, pp. 110–134) does not use the terminology of reflexive modernization, and environmental arguments are subordinate to his main theme, but his recent work on the information age and on the coming of the network society focuses on the same processes and interdependencies, directly relating several of the above-mentioned ideas on the changing character of modernity to the new trends and processes of globalization.

I have now given a brief overview of various interpretations of globalization—particularly since the 1970s, when the term 'globalization' was still unfamiliar. I have also provided a specific and restricted interpretation of what this volume considers globalization to be, and positioned this interpretation of globalization in the ongoing theoretical debate on the changing character of modernity. In this until now rather theoretical debate and analysis, the environment has appeared only incidentally, mainly because the environment features marginally in globalization theories and studies.

Before turning our attention fully to the environment in chapters 3–5 and beyond, we need to turn from this rather abstract overview of globalization perspectives and interpretations to a more substantive analysis of globalization as such. Globalization should not be portrayed merely as a theoretical concept. It is involved in actual transformations in the social practices and institutional forms of modern societies, and it plays a significant role in daily politics. To substantiate the global processes of social change and the debate on the extent and historic uniqueness of these transformations, I now turn to the processes of economic globalization, to the subject of globalization and culture, and to the political dimensions of globalization.

Globalization in the Economy: Global Capitalism

Globalization *avant-la-lettre* (at that time often referred to as internationalization or transnationalization) originated in the economic sphere of production and trade, and it was in that sphere that the debate on globalization actually began in earnest. The development of transportation and communication networks, the rapid growth of trade, the emergence of large transnational companies (Dicken 1997, 1998), and the huge flow of capital (mainly in the form of foreign direct investment) in the late nineteenth century all contributed strongly to the internationalization of the capitalist economy. Although international trade began earlier, as Wallerstein (1974, 1980) and Braudel (1979) have shown in fine detail, it was on the eve of the twentieth century that Marx so strongly emphasized the international character of capitalism. Accordingly, Hoogvelt (1997) has argued that, although both global trade and foreign direct investments are often presented as indicators of the recent era of globalization, their development and rapid increase over the last few decades should be seen as a "continuation" of a trend that began in the late nineteenth century. The geographic redirection of investment flows away from peripheral regions since the colonial period would in fact seem to indicate the opposite of globalization.[14] Hirst and Thompson (1996) also note that the growing dominance of transnational corporations in world production and trade shows historic continuity rather than dramatic change in the past 30 years, and therefore they reject the idea that a new world economic system is replacing the old international one. Both the share of world products subject to transnational corporate control and the relative number of international businesses in OECD countries still basically operating in their home territory (in terms of location of sales, affiliates, declared profits, research, and finance) are remarkably constant. National economies and nation-states, rather than a world economic system with transnational institutions articulated on a global scale, are important entities in the present-day economic world. Indeed, Hirst and Thompson (1996) conclude that if there exists such a thing as globalization then it is certainly nothing new—not a qualitative change, but merely a continuation of the international trading systems that paralleled the rise of capitalism, coming close to Robertson's definition of globalization.[15] Ruigrok and van Tulder (1995), Hoogvelt (1997), Dicken (1998), and others[16] are not so extreme in their conclusions. Nevertheless, they converge in putting into perspective both the global or "footloose" character of large corporations and the truly global character of production and technology; thus, they balance the ideas of Ohmae (1995) and other hyper-globalization theorists, who believe that nation-states and national economies are outmoded and disappearing institutions. The former group of scholars also emphasizes the fact that foreign direct investment and multinational enterprises are concentrated in three economic regions: the North American Free Trade Agreement (NAFTA) area, the EU region, and Japan and Southeast and East Asia. In that sense, globalization is limited to that very triad.

Geographic Concentration of the Global Economy: The Triad
When international flows throughout the world economy consisted primarily of raw materials, natural resources, and agricultural products, the spatial pattern of economic transactions was determined largely by differences in local natural circumstances. This meant that many countries now called "developing" were then important in the world economy. By the time industrial products, investment capital, finance, and specialized services had come to dominate global transactions, a new spatial pattern, no longer directly related to natural circumstances, had emerged. It was at that time that "triadization" began to emerge, although over time its specific form and appearance have depended on numerous economic, political, and other factors and have included a large contingency. Several indicators of the global economy point to the central position of the three constituent parts of the economic triad: the European Union, the NAFTA area, and Japan and the newly industrializing economies of Southeast and East Asia.

Transnational companies are still the major indicators of economic globalization. Various studies point at the fact that only 63,000 parent transnational companies, controlling about 690,000 foreign affiliates (UNCTAD 2000), now account for up to one-third of world output, 80 percent of global investment, and two-thirds of world export trade in goods and services. According to Sachs et al. (1998, p. 160), "the planet's 20 largest firms together have a greater turnover than the 80 poorest countries, and each of them earns more than even an entire state, like Malaysia, on the point of breaking through to modernity." Ruigrok and Van Tulder (1995), Dicken (1998, pp. 194–195), and UNCTAD (2000) have pointed out that all but one of the top 100 transnational companies (in terms of turnover or contribution to world economic development) have headquarters in a limited number of developed countries in the triad. Petroleas de Venezuela SA is the only non-triad-based transnational corporation in the top 100. The explosion of international (cross-border) mergers, takeovers, and acquisitions since the 1980s shows that these major economic actors in globalization have increasingly taken on a global aspect. The Lisbon Group (1995, p. 71) points out that 90 percent of the more than 4200 strategic intercompany cooperation agreements signed by enterprises around the world during the 1980s involved enterprises based in Japan, North America, and the European Union countries. Figure 2.1 illustrates the increase in cross-border mergers and acquisitions between 1987 and 1999.

Foreign direct investment (FDI) is often seen as a major driving force of globalization and as one of the main indicators of its "material" dimensions.[17] The tremendous growth in foreign direct investments over the past 20 years—a sevenfold increase (at constant prices), according to the OECD (1998)—is evidence of economic globalization. Deregulation, demonopolization, privatization, and the reform of trade and foreign investment regimes are believed to be central to the enormous increases in the levels of international direct investment in the 1990s. However, FDI is not distributed equally around the world. It originates primarily in the triad,[18] and it lands primarily in the developed countries in the triad and much less in developing countries outside the triad. Whereas in 1938 only 35 percent of the world's foreign direct investment was invested in the triad, by the mid 1980s (and also in the period 1995–1997; see Dicken 1998 and OECD 1998) about 75 percent of all FDI stock was in those countries. Sassen (1992) reports that in the 1980s FDI flows to and from developed countries grew at an annual rate of 24 percent, and that these percentages maintained similar levels in the period 1992–1997 (with a decrease in 1990–1992). However, the share of world FDI coming from developing countries has increased over the past 20 years (figure 2.2).[19] The emerging "tiger" economies of Asia, in particular, have managed to attract increasing amounts of FDI.[20] Among developing regions, Southeast and East Asia had taken over the lead from Latin America by the early 1990s, although the 1997–98 Asian economic crisis did have an adverse effect on FDI inflow (OECD 1998). Transnational company involvement occurred particularly in the more technologically advanced sectors (fine chemicals, electronics, computers), in the large-volume medium-technology consumer goods

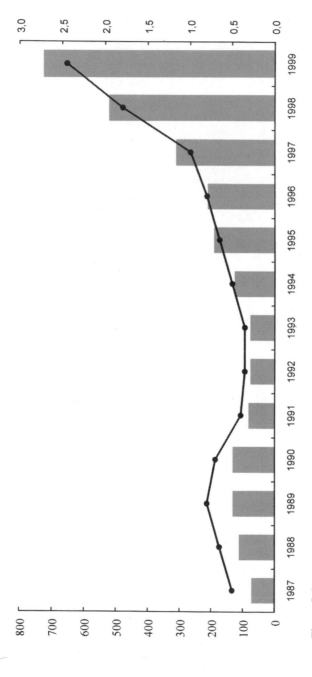

Figure 2.1
Value of cross-border mergers and acquisitions (M&As in billion dollars) and its share in GDP (percent), 1987–1999. Histogram bars: cross-border M&As. Line: share of cross-border M&As in GDP. Source: UNCTAD 2000.

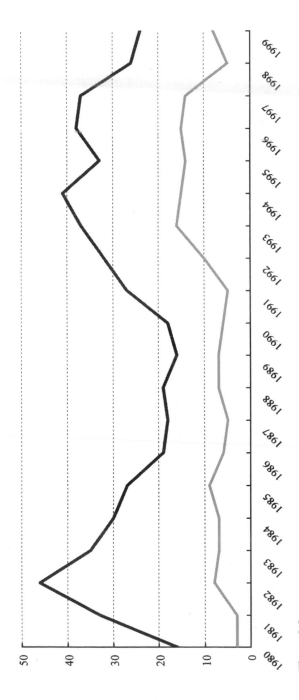

Figure 2.2
Share (percent) of developing countries in world flows of foreign direct investment (FDI), 1980–1999. Solid line: FDI inflows. Light line: FDI outflows. Source: UNCTAD 2000.

industry (cars, televisions, tires), and in industries supplying branded mass-produced consumer goods such as cigarettes and soft drinks (Dicken 1998). A recent innovation is the increase in small and medium-size industries engaged in FDI. This is beginning to challenge the traditional notion that only large companies can engage in foreign investment (Gentry 1999).

A third indicator of concentrated economic globalization is the development in trade flows, for example between 1970 and 1990. In the 1970s, 61 percent of world trade took place within and between the developed countries of the triad and some 14 percent completely outside it. In 1990, 75 percent of the world trade was within the triad; less than 5 percent was within or between the non-triad regions. (The remaining percentage represents trade between triad and developing countries.) Hoogvelt (1996), the Lisbon Group (1995), Castells (1996), Held et al. (1999), and others report similar trends and give more detailed data.

That similar developments can be seen in the world of financial products and services is a fourth indicator. The diverse service sector is important for all economies, developed and developing, in contrast to the conventional wisdom that it flourishes more in developed economies. However, it is particularly business services, related to the circulation of goods and financial capital, which have become more and more important in the global economy and which have concentrated in the triad (Dicken 1998, pp. 387–421). Nevertheless, as I will argue below, the relevance of these forms of economic globalization for environmental quality is rather indirect, and therefore I will not explore this issue further.[21] The overall pattern seems clear, however: In the 1980s and the 1990s, economic globalization was dominated by the prominent economies of the triad, while some up-and-coming "tiger" economies, especially in East and Southeast Asia, were increasingly becoming parts of that economic globalization. (Table 2.1 summarizes economic concentration in the triad.)

"Material" Dimensions of Globalization

The notion of globalization as triadization (or the concentration of globalization) is the starting point in refining the idea of globalization as a uniform economic process that stretches around the world, affecting all places equally. On this point Dicken, Castells, and the Group of Lisbon do not differ with Hoogvelt, Hirst and Thompson, and Ruigrok and van Tulder, who have focused on putting into perspective the truly global character of

Table 2.1
The global triad's concentration of manufacturing production and exports (in millions of US dollars for 1998, including inter-EU trade). Sources: WTO 2000, UNIDO 2000.

	Manufacturing value added (MVA)		Exports	
	Total	Percentage of world MVA	Total	Percentage of world exports
EU	1,731,700	30	2,176,000	38.8
NAFTA	1,642,800	29	1,069,700	19
Japan, Southeast Asia	1,644,400	29	1,186,100	24.3
Total	5,018,900	88	4,531,800	82.1

economic practices. Two additional clarifications bring the different perspectives of the two groups closer together, without diluting them. A first partial clarification is provided by the debate over homogenization and heterogenization. In the economic sphere, the trends toward universalization, standardization, and homogenization—usually seen as markers of globalization processes—are challenged by the need to diversify, individualize, and localize the economic products and processes of global capitalism. Though manufacturing, trade, services, and finances may increasingly be propelled by globalization, there are very rational foundations supporting the observed movements toward diversification, regionalization, and geographical concentration of these economic practices and processes. For instance, the major global actors—the transnational companies—show both universalizing and diverging characteristics in their structure and operations because the milieu of their native country continues to influence the development of their global operations. To an extent, American companies still operate differently in the global market than German and Dutch companies.

Waters (1995) supplies a second partial clarification of the diverging perspectives in globalization studies by distinguishing between different categories of economic processes. He concludes that material economic relationships (labor, manufacturing) have a tendency to localize, while symbolic relationships (financial markets, ideological arenas such as management concepts of flexibility and, increasingly, trade and investments in services) tend to globalize. It is mainly the latter forms of capital circulation, via modern communication and information technologies and along routes of increasing distance and growing velocity, that are at the root of the present phase of globalization and the new global economic regime (Sassen 1994). Hoogvelt (1997) adds to this analysis, observing that it is the integration of the global financial markets and especially "global financial deepening" that constitutes a truly "globalizing" force. This financial deepening occurs when the rate of growth of international financial transactions is much more rapid than any of the underlying economic fundamentals of the "real" economy (such as trade, investments, or outputs). In line with the above-mentioned decline of the Third World's participation in the fundamentals of the world economy, its share of total international bank lending has declined to about 11 percent, although at the same time the Third World elites have forcefully entered the global financial transaction market. This is only one side of a more general trend toward what Waters (1995) calls the "cultural economy" and Lash and Urry (1994) designate as the "economies of signs and space," in which markets involving "symbolic" goods have moved beyond the declining capacity of nation-states to manage them, while the economy is concurrently becoming increasingly subordinated to individual tastes and choices. The leading sectors in this transformation are those that produce the symbols: the mass media, the entertainment industry, and the service sector.

This, however, does not mean that manufacturing—often of more direct interest for environmental deterioration and improvement—remains untouched by the dynamics of economic globalization. Manufacture is also transformed by processes of liberalization, economic rationalization, privatization, deregulation, and economic restructuring, just as economic practices and institutions in the "symbolic" sectors are. Hoogvelt (1997, pp. 122–131) sees three main consequences of globalization for the "real" economy. The first is the emergence of one global market principle or "discipline" governing the world economy. We witness the development of an internationally integrated production system, embracing complex networks of inter-product and intra-product trade, and resulting finally in one dominant and globally set standard for price, quality, and efficiency. This market principle has two features. To start with, it not only directs international production and trade of services and goods, but also the national or domestic supply and production of consumer goods, intermediate goods, technology, raw material, labor, and capital. In addition, these standards are internalized by individual actors who are aware of global competition and hence feel compelled to conform to these international standards. In other words, these standards do not have to be enforced as they are reproduced via everyday experience and interaction (often linking activities many miles apart), media coverage, and interpretations of experts. It is the worldwide circulation of information that contributes so strongly to the reproduction and internalization of this principle or "discipline." The second consequence of globalization for the "real" economy relates to the delocalization of the production of goods and services. Although industrial production and other "real" economy activities (in contrast with international financial transactions) usually have some locational profile, the costs of transporting products and of transporting information have decreased to a level where routine production can increasingly be located anywhere and is thus in global competition. Of course this is easier for service production, such as data processing, but it is also beginning to simplify subcontracting in manufacturing. Here I touch upon new flexible, post-Fordist organization structures in and between multinational and domestic companies: loosely confederated transboundary networks or industrial complexes of production units. Manufacturing processes include the assembling of parts, produced in different locations all over the world by different firms, into product variations suitable for specific markets. The combination of computer technology with telecommunication has contributed significantly to these organizational innovations. The third consequence for the "real" economy relates to so-called global financial deepening. Global financial flows are the ultimate form of disembedding of capital from the (national) social relationships to which it used to relate. The free flow of financial capital at high speed around the stock markets of the core[22] of the global economy and the deterritorization of profits made by these flows also affect the real economy. There is no such thing as a free lunch: someone has to pay for the profits made by financial transactions, and according to Hoogvelt it is usually peasants and workers in the less developed regions who pay the price. The deterritorization of profits means that pension funds and the like do not have to wait until financial investments are turned into industries or skyscrapers; they can move their capital once profits go down, thus undermining national social solidarity. Economic crises initially caused by the financial markets have had an impact on the "real" economy, as the recent 1997–98 Southeast Asian financial crisis clearly illustrated.

To summarize: While economic globalization in manufacturing, in particular, should be put into perspective, and is still far from being a global network of footloose economic agents, tendencies toward globalization certainly affect manufacturing. It would be mistaken, therefore, to either view (the consequences of) globalization as restricted only to finances and services or designate globalization as a mere continuation of an international trade regime that started in the sixteenth century. In assessing the environmental consequences of globalization we have to realize that it is not the mere quantities or locational characteristics of services, goods, and capital globally transported or produced that are decisive. Modes of production, regulation styles, internationalization of competition, preferential treatment by national or regional governments, investment patterns that are increasingly beyond national control, technological innovations and diffusions, interference of local communities and NGOs, and other factors all contribute to the final environmental outcome of the processes of globalization. Therefore, the dynamics of these latter factors—articulated at all levels—need to remain very much in our minds in our attempt to analyze how the environment relates to economic globalization.

Global Culture: Between Global Producers/Local Interpreters and Local Producers/Global Interpreters

The most direct and strongest chronological line between globalization studies and the discourse on modernity runs via the typical postmodern emphasis on culture. Although several authors have tried to come to grips with the phenomenon of postmodernity through an analysis of the current and changing condition of global capitalism (Harvey 1989; Lash and Urry 1994), it should not surprise us that a large number of contributions to globalization theories (e.g., those of Robertson and Featherstone) still have a strong cultural emphasis. The principal issue in most of these contributions is the changing cultural institutions that are both the medium and the outcome of the processes of globalization. The globalization of culture on a transnational and transsocietal level takes various forms (religion, tourism, mass media, universal norms, eating practices, civil society movements) and is, in a number of respects, closely interrelated with global economic exchanges. Because of this some scholars link economic globalization (usually closely associated with capitalist modes of production) with an emerging global culture in its various forms and aspects. Originating from postmodernism, these cultural conceptualizations of globalization initially challenged the ideas of homogenization, universalism, and Westernization, which dominated the industrialism, capitalism, and rationalism brands of modernization theories in the 1960s and the 1970s.

Albrow (1997) distinguishes four dimensions of globalization that have been the focus of cultural studies. The first centers on global values that are drawn from the world and made to apply locally, such as eating habits, freedom of religion, and women's emancipation. This line of analysis also incorporates environmental norms and values that have become increasingly universal but time and again have to be made to work locally. A second tradition of studies focuses on local cultures that have access to and are influenced by events elsewhere. Transformations in these local cultures and resistance to interventions from outside are among the core fields of interest. A third branch of cultural studies concentrates on direct interactions between distinct parts of the world via telecommunications and the Internet—international NGO networks and their campaigns to stop the MAI are a recent example. Finally, some scholars study the maintenance of "traditional" lifestyles and life routines in new places, for instance by migrants.

These culture-oriented studies have yielded three central findings that are relevant to the subject of this book. First, people can live in one geographical place and have meaningful social interactions almost entirely outside that location, around the world. This aspect falls within all definitions of globalization, as it provides evidence of the lifting of social relations out of their local context and their regrouping across time and space. Second, the idea of the subordination of culture to the structural developments in, especially, the (global) economic-technological sphere is no longer seen as an adequate conceptualization. Instead, the emphasis has shifted toward diversity, variety, and local interpretations of the global mass media messages and international cultural phenomena that were so often interpreted solely in terms of global capitalism. The relative power of consumers' individual preferences and tastes, and the limited ability of powerful global economic producers and their advertising campaigns to manipulate market demand are often highlighted, thereby challenging the more traditional neo-Marxist

and World-System theses on the capitalist class order and transnational corporations. This can be summarized in the one-liner that global cultural producers are always dependent on local interpreters for the design and
marketing of their products. Third, the variety of local, traditional, reli-
gious, and other protests against global—often perceived as Western—cultural signs, tokens, and economic processes and activities sometimes result
in the recombination of local ideas and practices into global norms and
institutions. In a number of cases—among which environmental issues are
prominent examples—the locally oriented countermovements against some
strands of globalization have contributed to (cultural) globalization by
"upgrading" local ideas and actions into globally shared norms and values, global networks and organizations, and global exchanges of information. Not surprisingly, studies on international environmental relations and
regimes, which traditionally started from a political analysis, nowadays
emphasize the vital role of civil society and cultural dimensions in understanding global politics (Lipschutz 1996; Young 1997b; Wapner 1997). The
originally local interpreters of global culture can now be seen as local producers of a different kind of global culture; while the global producers of
cultural signs and tokens (the culture industry) increasingly have to interpret the cultural signs coming from local movements in order to function
successfully.

Political Globalization: From International Relations to Global Governance

The work of international relations theorists in particular—for example,
the (neo)realist school of thought and the (new or neo-liberal) institutionalists—has paved the way for political analyses of globalization. (Neo)real-
ists have devoted little if any attention to environmental issues in
internationalization, but institutionalists have recently "discovered" the
environment as an important field for international and global institution
building (Haas 1990; Young 1989, 1994, 1997a; Haas et al. 1993; Keohane
and Levy 1996). Most of these studies focus on the various state and,
increasingly, non-state[23] actors involved in international or global environmental politics, analyzing and identifying the basic constraints and opportunities for effective policy negotiations, institution and regime building,

and policy implementation. In spite of their major contributions to our understanding of international environmental regime formation and effectiveness, most of these studies concentrate on "piecemeal or issue-specific arrangements" (Young 1997b, p. 274) and remain far removed from a more overall sociological perspective on political globalization in relation to declining sovereignty and emerging global governance.[24]

The work on world polity by Thomas and colleagues (Thomas 1987; Boli and Thomas 1997), which combines cultural theories with political perspectives, is of some interest regarding this latter topic, although it touches only marginally on environmental issues. (For a recent environmental analysis in this tradition see Frank et al. 2000.) These institutionalists—not in the strict sense common in international relations theory, but rather in the broader sociological sense—emphasize the construction of a so-called world polity that is relatively independent of and irreducible to the world economic system (Meyer 1987). World polity is not merely the sum of traditional political categories such as states, transnational companies, or national forces, but includes transnational cultural and institutional frameworks.[25] Criticizing the inadequacies and shortcomings of a purely or mainly economic analysis of (production and exchange) institutions and relationships in the world system, Meyer (1987) and Boli and Thomas (1997) assert the importance of including global political and cultural frameworks in order to understand major globalization developments. These include transnational and global political and social institutions, global political and social actors, and the emergence of global and (increasingly) universal norms and values. They interpret the development of world polity as a process of political rationalization in a Weberian sense (distinct and separate from economic rationalization), which develops through conflict and struggle. In their analyses, they emphasize the universalizing tendencies of world polity and culture (which result in collective identities and interests of companies, societal organizations, states, and nations); they downplay heterogenizing and diverging developments. Although they conclude that, at the moment, world polity can be characterized as a world proto-state (lacking a single authority structure, but with shared cultural categories and shared principles of authority that are functioning), it may eventually develop into a world state or a world government—a conclusion not shared by everyone as I will illustrate below.

Declining Sovereignty

Although the world polity analysis concentrates on universalizing tendencies at the transnational level, another challenging—and complementary—line of argument within the discourse on global politics discusses the changing bureaucratic and political structures under the aspect of declining sovereignty. Although Castells fiercely opposes the idea that states no longer have any major economic role to play in an age of globalization and deregulation, he has to conclude (1997a, p. 354) that "bypassed by global networks of wealth, power and information, the modern nation-state has lost much of its sovereignty. By trying to intervene strategically in this global scene the state loses capacity to represent its territorially rooted constituencies." Although most decisions are still taken by national governments, and international organizations are merely a reflexive recognition and reinforcement of national sovereignty, "in a world of regional and global interconnectedness, there are major questions to be put about the coherence, viability and accountability of national decision-making entities themselves" (Held 1995). According to Held, sovereignty[26] is undermined by international law, the internationalization of decision making via regimes and organizations, hegemonic powers and international security structures, national identities, the globalization of culture, and economic globalization. I will briefly elaborate on the internationalization of decision making and economic globalization in relation to declining sovereignty, before I take up the question of global governance.

International politics can no longer be interpreted as merely the logical outcome of the international system of nation-states, since multilateral organizations and global regimes have attained their own internal logic, interests, and dynamics and have run partly "out of control" of the system of nation-states. Even hegemonic powers are to some extent "trapped" in the cage of global regimes and organizations, albeit to a lesser extent than the "average" nation-state. The diminishing political role of the nation-state due to the emergence of new international forms of political authority, is emphasized by McMichael (1996a, p. 39):

Internationalization of political authority includes both the centralization of power in multilateral institutions to set global rules and the internalization of those rules in national policy-making. . . . The definition of an international regime is thereby refined to include the actual determination, or at least implementation, of those rules by global agencies. In other words, the potential global regime is only formally multilateral, as states lose capacity as sovereign rule makers.

The participation and contribution of non-state actors—multilateral as well as sub-national—in regime alteration does not signify the end of the state in global politics, but contributes to a significant change in the state's role and primacy in processes of political globalization. This goes together with the growing belief that states should be held responsible, and even sometimes accountable, for the external effects of their own actions and also of the actions of various parties operating within their jurisdiction. This causes problems for nation-states under conditions of economic globalization, as transnational companies operate beyond national boundaries.

Challenging the conventional position of the nation-state at both the global and the national level, economic globalization has become highly significant. Some scholars (e.g., those of the vanishing Realist school) cling to the idea that sovereignty remains unaffected in the system of nation-states. At the same time, national political elites and states are often not very willing to give up state sovereignty and the powers connected with it. However, it is no longer a subject of much discussion that economic "globalization ruptures the territoriality of conventional international relations" (Williams 1995), that the state is less able to dictate or largely control the national developments in the economic, environmental, or other spheres, or that the growing contribution of private non-state actors in global politics undermines the central position of the nation-state. Still, in various policy domains affected by economic globalization the national state seems to have strengthened its position by, for instance, actively designing and implementing national economic and industrial policies. In her analysis of the consequences of economic globalization for the changing role of the nation-state, Hoogvelt (1997, pp. 134–139) concludes that much of the dispute between those who hold "declinist" views of the nation-state and those who claim to observe a strengthening of national authority can be attributed to the fact that national governments adjust their economies to globalization by, among other things, introducing policies for deregulation and privatization. Nation-states tailor their economies to fit the requirements of global competitiveness because they are forced to do so by the transnational business culture and ideology, by transnational and international (economic) institutions and actors (e.g., the structural adjustment programs of the IMF), and by "elite interactions."[27] They do this by actively gearing their national economies toward further privatization of the state sector, deregulation of monetary, fiscal and social policies, and restructuring of the welfare state. The growing importance of economic globalization does not necessarily result in a decline in state activity, but it does change the direction and the content of state performance.

Global Governance

Although there is some basic agreement on declining national sovereignty as a result of, among other things, economic globalization and the internationalization of political authority, the normative standpoints on the possibilities and desirability of (democratic) governance beyond the level of the nation-state are cause for argument among scholars. Held (1995, p. 236) and the world polity theorists make a strong case for a cosmopolitan or global model of democracy and governance.[28] Numerous environmental advocates join them in their demand for global environmental management institutions and structures. Porter and Brown (1991, pp. 143–156), for instance, identify three basic strategies for strengthening supra-national politics to confront global environmental problems: incremental change of international policy making via multilateral environmental agreements on specific topics, a global bargaining or partnership approach resulting in major shifts in policies of industrialized and developing countries, and the creation of new global institutions for environmental governance. Porter and Brown celebrate the last strategy, and Esty (1994), the Group of Lisbon (1995), the Commission on Global Governance (1995), and others elaborate on what this should mean in the field of global environmental regulation on trade and investment issues. Others are less convinced of either the possibility or the desirability of global, supra-national institutions. Although transnational corporations are often believed to be proponents of global regimes, Murphy (1994, p. 57) argues that transnational corporations might oppose a regulated "world without borders," since integrated world governance would be able to set common rules and standards which would diminish the capacity of transnational corporations to play off one nation or group against another. Others see proposals for developing a "world government" as naive insofar as they ignore the existing power structures and interests of established institutions.

Some of the causes for such skepticism are to be found in economic globalization processes, which both subvert the powers of the state at the national level and redefine the relationship between political entities and markets at the supra-national level. At the national level, the undermining of state sovereignty or autonomy results in shrinking capabilities for national policy making. The political room for maneuvering and the possibilities of the state's intervening against a narrow economic growth logic are seriously reduced, resulting in changing national governance models with a state that is less strong This is particularly true for policy areas that have a clear global outlook: both economic and environmental policy making are confronted by new models of national governance and politics. In the global arena, political entities at all levels (sub-national, national, and regional[29]) have become increasingly involved in economic competition against one another in order to attract international investment projects. Nation-states and other political entities have taken on some of the characteristics of businesses in their struggle for the "competitive advantage of nations" (Porter 1990). Political coordination on a global level is seriously jeopardized by these economic struggles of political entities.

There is fairly general agreement that the rise of global economic relations and plural authority structures alongside the nation-state does affect its sovereignty, although it has not (yet) resulted in a major withering away of the nation-state. Political and economic globalization bring about a redefinition of the relationship between nation-states and global political and economic institutions and non-nation-state authority structures and actors (transnational or sub-national); they also bring about a transformation of national policy-making styles and practices, because the nation-state's freedom to act is affected and it has to take various other actors and factors seriously into account in drawing up national programs.

3

Ecological Modernization: From National Emergence to Global Maturation

Reviewing the maturation of the environmental debate within the social sciences in the last decade, we seem to run into a contradiction.

On one hand, the majority of academic scholars, in rapidly and superficially assessing the environmental consequences of globalization, would easily agree on two conclusions: that the environmental consequences of globalization have attracted only limited attention among academics whereas environmentalists have increasingly taken a stand on this issue and included it in their campaigns, and that scholars and environmentalists who do address the two-way relationship between globalization processes and the environment tend to emphasize the negative effects of globalization on environmental quality. "The apocalyptic horizon of environmental reform" (Mol and Spaargaren 1993) in the current environmental debate seems particularly conditioned by global change, and thus it prevails in academic studies and assessments of globalization and in the arena of environmental politics.

On the other hand, many environmental sociologists studying current developments in national environmental reform agree that the all-pessimistic or even apocalyptic interpretation of a capitalist-industrial society unable to reform itself along lines of sustainability—the view that was so dominant in the 1970s and the early 1980s—has melted into the air. Maxeiner and Miersch set the tone with their 1996 book Öko-Optimismus, in which they emphasized the successes of environmental reform and gave an optimistic outlook for sustainability in the twenty-first century. Numerous other studies focus on, outline, and contribute to the environment-induced institutional transformations that are actually taking place, or are in status nascendi, in contemporary Western societies, instead of merely highlighting the devastation brought about by their capitalist-industrial institutions. Dematerialization, factor 4 (or 10, or 20),[1] industrial ecology, design for the environment, environmental management and auditing, and technological shifts are a few of the dominant themes in the more recent literature. Also, this shift in the sociological perspective appears to coincide with a shift in the activities of environmental movements at the start of the new millennium (Mol 2000). Linking up with both national governmental agencies and the private sector, environmental NGOs are increasingly becoming involved in and contributing to concrete environmental reform.

How can we deal with these two positions in the environmental discourse? Should we interpret them as mutually exclusive extremes of a debate in which different environmental theories, ideas, perceptions, and ideologies clash, or can we assess their relevance and value for different practices and arenas, time frames, and geographical scales by analyzing their origins and backgrounds? I shall attempt to clarify this seeming contradiction between national perspectives on ongoing ecological reform and the apocalyptic environmental horizons of global change by first providing a historical overview. This will highlight the dominant analyses and perceptions of the social origins of and solutions to what might be called "the environmental question." The environmental discourse on causes and solutions has been through different phases, each centering on distinctive institutional traits that were either held responsible for environmental deterioration or seen as crucial for structural and lasting environmental reform. I will then discuss two recent innovations in the environmental debate that are closely related to the above-mentioned contradiction: the establishment of the notion of global change or globalization and the emergence and substantiation of ideas of ecological modernization. The relevance of the theory of ecological modernization for some industrialized areas of the triad has been acknowledged, and I will therefore especially explore the question of its suitability for other parts of our globalizing world.

The History of a Changing Environmental Discourse

The history of environmental concern in Western industrial societies is usually divided into two or three different waves, depending on the historical outlook of the authors (table 3.1). At the outset of the twentieth century, a first wave of environmental concern could be discerned in almost all industrializing countries. This concern focused mainly on the degradation of "natural" landscapes due to increasing industrialization and the expansion

Table 3.1
Some characteristics of the three waves of environmental concern and reform.

	First wave	Second wave	Third wave
Beginning	Ca. 1900	Ca. 1970	Late 1980s
Central notion	Nature conservation	Limits to growth	Global change
Focal point	Protection of reserves and species	Minimizing additions and withdrawals	Sustainable development
Geographical range	Industrializing nation-states	Industrialized nation-states	Globalizing world
Results	Protected areas and species	National environmental agencies, laws, NGOs	Ecological reform of modern institutions around production and consumption
Major social theories on environment	—	Deindustrialization, neo-Marxism	Ecological modernization

of cities,[2] and was largely limited to some well-educated segments of society. Environmental concern did not lead these elites to question the foundations of the emerging industrial society. The emphasis was rather on demands for the protection of a limited number of endangered species (especially birds) and areas of great ecological importance against the devastating influence of rapid industrialization and urbanization. Nature reserves, parks, and (semi-)protected areas and species are the typical products of this wave in most industrial and industrializing societies. These "environmental reforms" were often initiated and triggered by private initiatives, to be taken over partially by the state, often after a considerable time.

The Second Wave

Most contemporary environmental sociologists, however, begin their analysis of environmental concern and social struggle for the environment with the second wave of environmental awareness, which came in the late 1960s and the early 1970s. Both the social carriers and the objectives of this second wave differed fundamentally from those of the first, as various environmental sociologists and historians have noted. The central notion of environmentalism in the 1970s was that a fundamental reorganization of the social order was a *conditio sine qua non* for an ecologically sound society. Although there was considerable debate as to which institutional characteristics should be held primarily responsible for the ongoing environmental deterioration, the promoters of this second wave converged on the conviction that radical changes would be necessary to alter the dominant trend. However, the ecology-inspired demand for radical social transformation in the early 1970s resounded only marginally in the institutional arrangements of industrial society. Among the most significant successes were the creation of government departments for the environment in most industrial societies (Jänicke 1990), expanding environmental legislation and planning, a growing number of international environmental organizations and treaties, and a rapid increase in the number and membership of non-governmental environmental organizations (figure 3.1). Again, these demands for change, as well as the social responses to and reactions against them, occurred almost simultaneously in most industrialized nations, including Japan (which took the lead in the 1960s), the United States, and the European triad countries. Many measures to combat environmental destruction were adopted and some were actually implemented. However, most of the challenged institutions of modernity, such as those that play major roles in the industrial structure, economic relations and scientific-technological developments, were not deterred from their devotion to a narrowly defined economic "progress." Ecological reform following the second wave did not affect the basic institutions that were held responsible for environmental disruption. In that sense, the second wave was unsuccessful, and its initiatives for environmental reform were often considered to be mere window dressing.

The meager results of institutional change in the 1970s and the 1980s are reflected in the dominant social theories on environmental degradation and (failing) environmental reform of that era. Neo-Marxist and "de-industrialization" or "counterproductivity" theories were predominant among the attempts to explain and understand the continuing cycle of environmental crisis and stagnating environmental reform. Neo-Marxists, such as Hans Magnus Enzensberger (1974 [first published in 1973]), Allan Schnaiberg (1980), David Pepper (1984), and, more recently, Peter Dickens, Ted Benton, and James O'Connor, emphasized the central role of the capitalist mode of production as both the cause of environmental degradation and the obstacle to overcoming it. De-industrialists or counterproductivity theorists, such

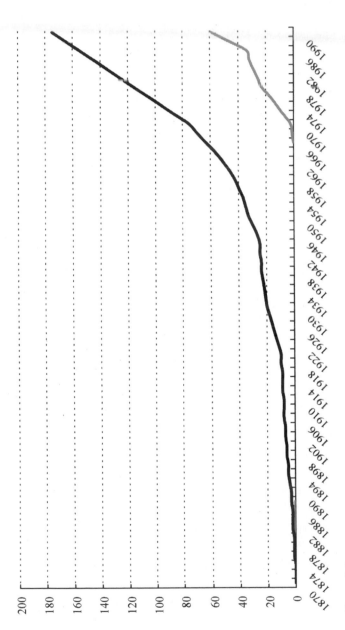

Figure 3.1
Cumulative numbers of national environmental ministries (light line) and international environmental non-governmental associations (solid line), 1870–1990. Source: Meyer et al. 1997, p. 625.

as Murray Bookchin (1980), Rudolf Bahro (1977, 1984), Otto Ullrich (1979), and André Gorz (1989 [1988]), concentrated primarily on the devastating influence of large technological-industrial developments (often, but not always, in relation to the capitalist mode of production). In that sense, the sociological debate in the 1960s as to whether industrialism or capitalism should be seen as the central characteristic of Western societies, continued into the 1970s in the environmental arena.[3]

In the 1970s the only exceptions to this censure of the failures of modern institutions to preserve the sustenance base were the post-industrial and post-materialist theories (Bell 1976; Inglehart 1990). Simultaneous with the discourse on the transformation of modernity and the emergence of "globalization" in the 1970s a radical environment-informed institutional reform was envisaged. The coming of the post-industrial or information society was supposed to result in both a dematerialization and a quantitative reduction of manufacturing and resource-intensive consumption, in favor of services, non-industrial sectors and non-material consumption patterns. This scenario of the "post-industrial utopians" (Frankel 1987) was not borne out by reality, as Giddens (1982), Kumar (1995), Castells (1996), and Dicken (1998) pointed out. The relative contribution of services, as compared to manufacturing and particularly agricultural or resource based sectors, to the national economies of most triad countries did increase in the 1970s and the 1980s. However, this did not lead to major environmental improvements due to diminishing industrial sectors, dematerialization or more environmentally friendly life styles and patterns of consumption.

The Third Wave

A third upsurge of concern for the "burdening of the sustenance base" in industrial societies first became noticeable in the late 1980s.[4] The Brundtland report (WCED 1987) and the 1992 UNCED conference in Rio de Janeiro are often cited as the markers and milestones of this third wave. They are the equivalents of the Report to the Club of Rome and the 1972 Stockholm conference, which played such important roles during the second wave. Some important distinctions have been observed in comparisons between the environmental upsurge in the 1970s and the third rise in environmental awareness in the late 1980s and the early 1990s. For instance, the leaders of the third upsurge were found among a wide variety of classes in Western societies, and even in developing countries. Most commentators agree on the differences between the second and third waves of environmental concern, just as environmental commentators noticed the difference between the second wave and the first one occurring during the late nineteenth and early twentieth century.

With a view to my line of argument I wish to emphasize two major distinctions or innovations. First, the nature of the contemporary environmental debate and the notions employed in it differ considerably from the debate and the notions under debate in the 1960s and the 1970s. One of the most prominent changes has been the emergence of "global change." Second, as I and my colleagues have argued extensively elsewhere (Spaargaren and Mol 1992; Mol and Spaargaren 1993, 2000a,b; Mol 1995, 1996; Spaargaren 1997; Spaargaren, Mol, and Buttel 2000), there is a crucial difference in that in the 1970s institutional transformations aimed at the preservation of the sustenance base were often no more than wishful thinking, whereas the last decade of the twentieth century saw the commencement of actual environment-induced transformations of the institutional order of modernity. Today, the institutional transformations to protect the environment that have been incepted can no longer be interpreted as mere window dressing. Others too have labeled these institutional transformation processes the "ecological modernization" of production and consumption, and, although severely criticized,[5] these ideas have increasingly been taken up by social scientists, by environmental commentators, and by those involved in daily environmental struggles, politics, and policies. The Ecological Modernization Theory has made an attempt to explain not only the discursive changes, as emphasized by authors such as Albert Weale (1992) and Maarten Hajer (1995), but also the actual transformations in social practices and institutional developments. The Ecological Modernization Theory has thus created the theoretical basis for a growing confidence that environmental reform can no longer be viewed exclusively in apocalyptic terms.

The emergence of "global change" and "ecological modernization" ideas will be elaborated in the following two sections.

From Global Environmental Change to Globalization

In an early stage of the third wave of environmental concern, the American environmental sociologist Fred Buttel and his colleagues (1990) noticed that the central notion in the environmental debate had shifted from "limits to

growth" to "global change." This notion underlined the growing importance of those environmental problems that concern and challenge the
entire world. The considerable attention devoted to environmental problems of global dimensions and the reframing of environmental problems in
terms of global change were characteristic of the third wave of environmental concern. It was a notion that was firmly put on the public and political agendas for the first time by the Brundtland Commission. From the late
1980s on, the environmental agendas of triad nation-states and the international community seemed increasingly dominated by problems related
to so-called global commons, such as the greenhouse effect, ozone-layer
depletion, diminishing biodiversity, rapidly vanishing tropical forests, and
pollution of the oceans.

This transformation of the environmental debate is more than just the
emergence of a different set of more fashionable environmental topics, or a
reframing of existing environmental agendas in new environmental concepts.

These "new" environmental challenges—often brought together under
the heading of Global Environmental Change (GEC)—had a number of
implications:

• They strengthened the old idea of global ecological interconnectedness,
which had already been put forward during the 1970s. However, at that
time GEC had very little impact on official political agendas and strategies,
which were still basically oriented toward the institutions of the nation-
state. The emergence of these global environmental problems onto the public and political agendas enhanced the sense that we live on one earth, but
also in one world.

• Consequently, global environmental change created strong political links
between different parts of the world, especially between the developed
North and the developing South (which until then had not been regarded
as politically interdependent in dealing with environmental challenges).
Now, the triad and the "periphery" were increasingly seen as partners in
a "shotgun wedding" within but also outside the growing number of (environmental) international institutions. Numerous examples given in this
volume illustrate this, although most are of quite recent origin. This does
not imply that an efficient collaboration between North and South has
evolved easily or is about to evolve in the near future; according to most
commentators, conflict rather than collaboration still seems to characterize
global environmental politics.

• Many scholars agree that global environmental change has strengthened
globalization processes and perceptions. Inversely, economic globalization

dynamics (or global capitalism) are also seen to be responsible for triggering and increasing global environmental change, as well as significantly aggravating a variety of other environmental problems, from local resource depletion to the pesticides "circle of poison."

• It was primarily these global environmental challenges that inspired sociologists to place the environmental dimension centrally in their theories on the transformation of modernity. The specific features of the era of late, reflexive, or global modernity can be fully understood only by appreciating how society deals with global environmental change.

Although all four implications follow from the emergence of global environmental change onto the political agenda, they are not automatically and causally interconnected, and the contribution of each to the changing environmental discourse is far from similar or synergistic. The changing terms of the environmental debate demonstrate that any analysis of the new phase of environmental concern and reform cannot be limited to global environmental change (the greenhouse effect, the depletion of the ozone layer, the pollution of international oceans and the issue of biodiversity being the most prominent aspects). That is, in order to understand global change, environmental sociology and the other social sciences have to focus on the relationship between the various dimensions of globalization processes (political, socio-cultural) on the one hand, and all manifestations and distributions of environmental deterioration, environmental perception and environmental reform (and not only those related to global issues) on the other. Global environmental problems—or "high consequence risks," as Giddens (1990) calls them[6]—are of course interesting and challenging as starting points for an analysis of how globalization interferes with environmental quality, and are to some extent markers of the new phase of environmental concern and reform. However, it might distort our theoretical understanding and models and limit our contribution to general sociological theories on late, reflexive and global modernity if we were to limit the definition of global change in such a way (as, for instance, Wilenius (1999) seems to do). In fact, sociologists who limit their analysis of globalization to global environmental problems are trapped in a natural science definition of environmental problems. In order to avoid this trap I will therefore use the expression "environmental dimensions of globalization processes" (whether they are negative side effects, enhanced environmental reforms or changing environmental perceptions) instead of the catch phrase "global change."

The Ecological Modernization Theory

The second innovation that transpired in the third wave of environmental concern deals with actual environmental restructuring in the most developed countries. Empirical studies on ecological restructuring have focused on distinct levels of analysis: individual industries; industrial sectors, zones and networks; or even industrial countries (Jänicke et al. 1992; Jänicke 1995). These studies try to assess whether a reduction in the use of natural resources and/or the discharge of emissions can be identified from the early 1980s up to the mid 1990s, either in absolute or in relative terms, compared to economic indicators such as GNP. This development is manifest in the recent explosion of studies on cleaner production, on industrial metabolism or industrial ecology (Ayres and Simonis 1995; Fisher-Kowalski 1996),[7] on dematerialization, and on industrial transformation (HDP 1996; RMNO 1996).

Although not all the conclusions in these studies point in the same direction, the general picture can be summarized as follows. From the mid 1980s on, a rupture in the long established trend of parallel economic growth and increasing ecological disruption was identified in most of the ecologically advanced nations, such as Germany, Japan, the Netherlands, the United States, Sweden, and Denmark. This slowdown is often referred to as the decoupling or delinking of material flows from economic flows. In a number of cases (regarding countries and/or specific industrial sectors and/or specific environmental issues) environmental reform can even result in an absolute decline in the use of natural resources and discharge of emissions, regardless of economic growth in financial or material terms (product output).

The social dynamics behind these changes, that is the emergence of actual environment-induced transformations of institutions and social practices in industrialized societies, are encapsulated in the Ecological Modernization Theory. This theory tries to understand, interpret and conceptualize the nature, extent and dynamics of this transformation process.

Historical Development of Ecological Modernization Ideas

The development of the Ecological Modernization Theory began—primarily in a small group of Western European countries, most notably Germany, the Netherlands, and the United Kingdom—around 1980. Social scientists such as Joseph Huber, Martin Jänicke, Volker von Prittwitz, Udo

Simonis, and Klaus Zimmermann (Germany), Gert Spaargaren, Maarten Hajer, and Arthur Mol (Netherlands), and Albert Weale, Maurie Cohen, and Joseph Murphy (United Kingdom) have all made substantial contributions to the early or later stages of the development of the Ecological Modernization Theory. In addition, various empirical studies using this theoretical framework have been carried out in various other countries, among which are Finland (Jokinen and Koskinen 1998; Jokinen 2000; Sairinen 2000), Sweden (Lundqvist 2000), Canada (Harris 1996), Denmark (Andersen 1994), Europe (Gouldson and Murphy 1996; Neale 1997; Mol, Lauber, and Liefferink 2000), the United States (Pellow et al. 2000), Lithuania (Rinkevicius 1998 and 2000), Hungary (Gille 2000), Kenya (Frijns et al. 1997), and Southeast Asia (Frijns et al. 2000; Sonnenfeld 2000).

Throughout the relatively short time of its existence, there has been considerable diversity and internal debate among the various contributors to the Ecological Modernization Theory. These differences spring from national backgrounds (with authors referring to various empirical references and interpretations, as I will illustrate below) and theoretical roots,[8] but also chronology. Though an extensive analysis and overview of ecological modernization literature up to now is outside the scope of the present volume (for that, see Spaargaren et al. 2000 and Mol and Sonnenfeld 2000), I believe it makes sense to distinguish at least three stages in the development and maturation of the Ecological Modernization Theory. The first contributions, especially those by Joseph Huber (1982, 1985, 1991), were characterized by: a heavy emphasis on the role of technological innovations in bringing about environmental reforms, especially in the sphere of industrial production; a rather critical attitude toward the (bureaucratic and inefficient) state, as found in the early writings of Martin Jänicke (1986); a very optimistic, perhaps naive, attitude toward market actors and market dynamics in environmental reforms (later on glorified by neo-liberal scholars); a system-theoretical perspective with a relatively underdeveloped concept of human agency and social struggle; and a concentration on national or sub-national studies. Some of the more critical remarks on the Ecological Modernization Theory still refer to these initial contributions.[9]

Building upon several of these limitations, ecological modernization studies in the second period, from the late 1980s on, showed less emphasis on and a less deterministic view of technological innovations as the motor behind ecological modernization. These contributions gave evidence of a more balanced view of state and market dynamics in ecological transformation processes, as illustrated by the work of Albert Weale (1992) and the later Martin Jänicke (1991, 1993). During this phase, the institutional and cultural dynamics of ecological modernization were given more weight, as well as the role of human agency in environment-induced social transformations. The emphasis remained on national or comparative studies of industrial production in triad countries. Critical remarks on the concept of ecological modernization in this period—articulated by scholars both inside and outside the ecological modernization tradition (Mol 1995; Blowers 1997; Blühdorn 2000)—focused on its Eurocentrism, since the Ecological Modernization Theory had been developed primarily in the context of a small group of Western European countries. In addition, comments pointed out its limited definition of the environment (Spaargaren and Mol 1992; Mol 1995), its overly optimistic expectations of environmental reforms in social practices, institutional developments and environmental debates, and its disregard for lifestyles and consumption practices.

The third period, from the mid 1990s on, encompasses innovations in three fields. First, studies on industrial production were increasingly complemented by work done on ecological transformations related to consumption processes (Spaargaren 1997; Spaargaren and van Vliet 2000). Second, the Eurocentrism criticism of the second period resulted in various national studies on environmental reforms in non-EU countries (newly industrializing countries and the transitional economies in Central and Eastern Europe, but also, for instance, the United States and Canada), leading to mixed conclusions on the relevance of this theoretical framework for understanding the processes of environmental reform. Finally, growing attention was paid to the global dynamics of ecological modernization (Spaargaren, Mol, and Buttel 2000).

In spite of national, temporal and theoretical differences, all these contributions can still be gathered together under the umbrella of the Ecological Modernization Theory, not only because they identify themselves as such, but also because they have the following things in common:

• that environmental deterioration is conceived of as a challenge for socio-technical and economic reform, rather than the inevitable consequence of the current institutional structure

- an emphasis on the actuality and necessity of transformation of modern institutions in the fields of science and technology, the nation-state, and global politics, and the global market in order to achieve environmental reform
- a position in the academic field that is distinct from the rather strict neo-Marxists, as well as from counterproductivity and post-modernist analyses.

Against this shared background, I will outline the central theoretical notion behind the variety of contributions to the Ecological Modernization Theory, as well as its core features.

Fundamental Premises of Ecological Modernization

The basic premise of the Ecological Modernization Theory is the centripetal movement of ecological interests, ideas and considerations involved in social practices and institutional developments, which results in the constant ecological restructuring of modern societies. Ecological restructuring refers to the ecology-inspired and environment-induced processes of transformation and reform going on in the central institutions of modern society. Institutional restructuring is of course not a new phenomenon in modern societies, but a continuous process that has accelerated in the phase that is often labeled late, reflexive or global modernity. The present phase differs from the pre-1980s phase, however, in the increasing importance of environmental considerations among the triggers for these institutional transformation processes in industrial societies.

Within the Ecological Modernization Theory this process is conceptualized at an analytical level as the growing autonomy or independence of the ecological sphere and ecological rationality with respect to other spheres and rationalities (Mol 1995, 1996; Spaargaren 1997). As mentioned in section two of this chapter, some notable environment-informed changes in the domain of policies, politics and ideologies already took place in the 1970s and the early 1980s. The construction of governmental organizations and departments dealing with environmental issues dates from that era. A distinct green ideology—as manifested by, for instance, environmental NGOs and environmental periodicals—began to emerge in the 1970s. However, in the 1980s in particular this ideology assumed an independent status and could no longer be interpreted in terms of the old political ideologies of socialism, liberalism and conservatism. This new independence was underlined by authors such as Phaelke (1989) and

Giddens (1994a).[10] However, the crucial transformation which makes the notion of the growing autonomy of the ecological sphere and rationality especially relevant, is of more recent origin. After the ecological sphere and rationality have become relatively independent from the political and socio-ideological spheres and rationalities (as occurred in the 1970s and the 1980s), this process of growing independence has begun to extend to the economic sphere and economic rationality. And since, according to most scholars, this growing independence of the ecological rationality and sphere from their economic counterparts is crucial to "the ecological question," this last step is the decisive one. It means that economic processes of production and consumption are increasingly analyzed and judged, as well as designed and organized from both an economic *and* an ecological point of view. Some profound institutional changes in the economic domain of production and consumption have become discernible from the late 1980s on. Among these changes were the widespread emergence of environmental management systems, the introduction of an economic valuation of environmental goods via the introduction of eco-taxes, among other things, the emergence of environment-inspired liability and insurance arrangements, the increasing importance attached to environmental goals such as natural resource saving and recycling among public and private utility enterprises, and the articulation of environmental considerations in economic supply and demand. The fact that we analyze these transformations as *institutional* changes indicates their semi-permanent character. Although the process of ecology-induced transformation should not be interpreted as linear and irreversible, as was common in the modernization theories in the 1950s and the 1960s, these changes have some permanency and would be difficult to reverse.

Some environmental sociologists and commentators go one step further. They suggest that environmental considerations and interests not only activate institutional transformations in contemporary industrial societies, but even evolve into a new Grand Narrative (de Ruiter 1988). The traditional Grand Emancipatory Narratives of modernity (e.g., the emancipation of labor, the dissolution of poverty) place us in history as human beings who have a definite past and a somewhat predictable future. Now these narratives have ceased to perform as overarching "story lines," some believe the ecology will emerge as the new sensitizing concept through which industrial society orients itself in its future development. Environment-informed institutional transformations are among the core features of this new Grand Narrative, according to these environmental sociologists. The environment—or rather environmental considerations and interests—is then a prime candidate to form the core structuring principle, the leitmotiv for processes of social transformation and continuity in what can be labeled (in a variation on Hobsbawm) the "Age of Environment." Whether or not this is an exaggerated interpretation of existing *tendences lourdes* can only be concluded after some decades, since at the moment we seem to be only at the beginning of the process in which environmental considerations and interests assume a more profound role in social development. However, the mere suggestion of such a possibility moves us far away from any postmodern interpretation of environmental transformations.

Ecological Modernization as Environmental Restructuring

Based either explicitly or implicitly on these core theoretical premises, most ecological modernization studies focus on actual environment-induced transformations in social practices and institutions. As I have analyzed elsewhere (Mol 1995), the core features of these transformations can be grouped in five categories:

• The changing role of science and technology in environmental deterioration and reform: first, science and technology are not only judged for their role in causing environmental problems, but also valued for their actual and potential role in curing and preventing them; second, traditional curative and repair options are replaced by more preventive socio-technological approaches that incorporate environmental considerations from the design stage of technological and organizational innovations; finally, the growing uncertainty with regard to scientific and expert knowledge on the definitions of, causes of and solutions for environmental problems does not result in a denigration of the contributions of science and technology to environmental reform.

• The increasing importance of economic and market dynamics and economic agents: producers, customers, consumers, credit institutions, insurance companies, the utility sector, and business associations increasingly turn into social carriers of ecological restructuring, innovation and reform (in addition to, and not so much instead of, state agencies and new social movements; see Mol and Spaargaren 2001 and Mol 2000). This goes together with changing state-market relations in environmental reform.

• Various transformations regarding the traditional central role of the nation-state in environmental reform: first, there is a trend toward more decentralized, flexible and consensual styles of national governance, at the expense of top-down hierarchical command-and-control regulation (a trend often referred to as political modernization; see Jänicke 1993); second, greater involvement of non-state actors taking over the traditional tasks of the nation-state (privatization, conflict resolution by business-environmental NGO coalitions without state interference, and the emergence of subpolitics[11]); finally, an emerging role for international and supra-national institutions that to some extent undermine the traditional role of the nation-state in environmental reform.

• A modification of the position, role and ideology of social movements (with respect to the 1970s and the 1980s) in the process of ecological transformation: instead of positioning themselves on the periphery or even outside the central decision-making institutions on the basis of demodernization ideologies, environmental movements seem increasingly involved in decision-making processes within the state and, to a lesser extent, the market. This goes together with a bipolar or dualistic strategy of cooperation and conflict, and internal debates on the tensions that are a by-product of this duality (Mol 2000).

• Changing discursive practices and the emergence of new ideologies in political and societal arenas, where neither the fundamental counterpositioning of economic and environmental interests nor a total disregard for the importance of environmental considerations are accepted any longer as legitimate positions. Intergenerational solidarity in the interest of preserving the sustenance base seems to have emerged as the undisputed core and common principle.

Apart from the fact that these transformations do not occur in all studied countries in the same way, to the same extent and at the same time, these heuristics are also applied in different ways by different scholars. Some see them as analytical tools to be used in order to understand the emerging contemporary processes of environmental reform. Others view them also as guidelines for designing the best and most effective environmental reform models for the future.

Irrespective of the extent to which these environment-induced transformations actually will take place, the conclusion stands that the interpretation of what is happening during the third wave of environmental concern is reflected in the social theories that dominate the current environmental debate. Within the subdiscipline of environmental sociology (or at least the Western European contributions) neo-Marxist and counterproductivity theories have shifted somewhat into the background, while models and ideas along the lines of ecological modernization are on the rise.[12]

Ecological Modernization beyond the Triad: A Case of Neo-Colonialism?

As noted above, ecological modernization ideas and related ecological restructuring models have been (and to a major extent are still being) developed in the context of a limited number of developed societies in the triad. The socio-political, economic, and cultural conditions and institutions in this geographical area have played an important role as the background against which these ideas have been developed and on which the resulting reform models and practices have been successfully applied. Although scholars such as Albert Weale (1992) have deliberately included the international dimension in the "new politics of pollution," this has not been done in a systematic way for the majority of ecological modernization ideas and contributions.

This observation of what can be labeled "Eurocentrism" in the Ecological Modernization Theory leads to two interrelated questions that challenge ecological modernization ideas. Both are of particular relevance to this study on globalization and environmental reform. First, it has been questioned whether this Eurocentric cultural, political-institutional and economic bias conditions and confines the geographical area to which the Ecological Modernization Theory can be applied—in both a descriptive or analytical and a normative sense. Second, if the applicability of the Ecological Modernization Theory should prove to be geographically restricted, what would that mean in the context of a global modernity? In a globalizing world, it is still the modern industrialized countries that provide the dominant models for (economic) development. The concepts, institutions and practices of ecological reform developed in the triad are not necessarily appropriate for non-triad countries, but they might still be "imposed" upon those countries through a range of globalization mechanisms that tend toward harmonization. Or to put it a different way: What are the opportunities for non-triad countries in particular to pursue development paths that differ from the dominant Western models of environmental reform, which seem so strongly connected to the ideas of ecological modernization?

Ecological Modernization and Non-Triad Societies

The question of the suitability of the Ecological Modernization Theory for developing countries has prompted rather critical remarks, coming from both ecological modernization scholars themselves (Mol 1995; Frijns et al.

2000; Sonnenfeld 2000) and their commentators and critics (Blowers 1997; Blühdorn 2000; Buttel 2000b). They all agree that in some of these developing countries ecological modernization heuristics are at best of partial value in analyzing environmental reform processes and practices. Although the specific institutional conditions in modern industrial societies differ in a number of vital aspects from those in some industrializing societies, considerable disparity exists between the newly industrializing countries in East and Southeast Asia, Latin America, and the Central and East European countries in transition, on the one hand, and most developing countries in sub-Saharan Africa, on the other.[13]

With regard to the latter category in particular, the Ecological Modernization Theory has little analytical or descriptive value, since the (non)dynamics of environmental protection and reform seem to follow a different "logic." A "transfer" of the ideas of ecological modernization and related environmental reform institutions and practices to these countries and regions would therefore be similar to the transfer of "green revolution" technology from industrialized to developing countries. Both would imply an enforcement of Western ideas that are ill-adapted to local social, institutional, ecological and cultural contexts. Ideas and institutional reforms of ecological modernization are based on specific assumptions, which are often not valid for these developing regions: for example, the existence of a welfare state with articulated and institutionalized environmental responsibilities, advanced technological development within a highly industrialized society, a state regulated market economy covering all parts of society and closely connected to the globalized world market, and relatively profound, widespread and institutionalized environmental consciousness. In earlier writings (Mol 1995; Frijns et al. 2000), I and my colleages have argued that these institutional characteristics are preconditions for the successful application of environmental reform models based on the Ecological Modernization Theory.[14]

With respect to Central and Eastern European (CEE) countries, ecological modernization ideas seem to be more appropriate in their analytical and normative connotations. Although considerably different, at least a number of these countries have a relatively strong welfare state that is changing into a more communicative and less directive, top-down mode of operation (although the ministries for the environment are generally still weak), an emerging market economy, relatively sophisticated systems in the fields of (environmental) science and innovation, varying levels of democracy and public participation, but usually limited spread of environmental consciousness and relatively powerless environmental NGOs.[15] A similar line of argument—although with different qualifications on the environmental matching of their institutional order to fit ecological modernization reform models—is valid for the newly industrializing countries in, for instance, East and Southeast Asia (O'Connor 1994; Rock 1996a; Sonnenfeld 2000).

To conclude, ecological modernization ideas of environmental reform seem especially relevant for developed societies and—to an increasing extent—newly industrializing countries and some CEE countries. Chapter 8 analyses developments in environmental protection and reform in a number of non-triad countries in greater detail, specifically against the background of globalization.

Ecological Modernization in a Globalizing World

However—and this is the second question—ecological modernization has also been subject to more radical criticism from what can be called a Third World perspective. Social scientists involved in environment-and-development studies, such as Ted Trainer (1988), Vandana Shiva (1989), Saral Sarkar (1990), and Wolfgang Sachs (1993), have challenged the perspective of ecological modernization for *industrialized societies*.[16] Pointing at the interrelatedness and mutual interdependence of triad countries and developing countries in the era of globalization, these authors have advanced two interrelated arguments in criticizing ecological modernization. First, they claim that the theory is only applicable to industrialized societies, because it largely disregards the issue of an equal distribution of natural resources among different groups and nation-states. In line with World-System theorists and neo-Marxists, these scholars argue that an ecological modernization strategy in the most developed part of the world is only possible because of the net withdrawal of natural resources from and the net addition of pollution to developing countries by the more developed ones (Goldfrank et al. 1999). Agarwal and Narain (1991) refer to "environmental colonialism" in their analysis of global warming in the present global world, where the benefits and costs of major greenhouse gas emissions are unequally distributed between the rich triad countries and the poorer South. Second, under conditions of high levels of time-space compression and a dominance of industrialized nation-states and "their" transnational corporations, these authors are skeptical of the opportunities for developing societies to unfold their own strategies and models in dealing with their own environmental crises. Economic interdependence, intensifying global political interaction and collaboration, worldwide standardization of (Western) science, technology, production and consumption, and an emerging global culture create converging tendencies in environmental reform programs. Hardly any room is left for developing countries to choose their own "ecologically sound development path." This argument warns against the "enforced" introduction into developing countries of the reform model of ecological modernization, since it was not "designed" for the social structures and institutions prevailing in developing countries. These critics assert that, exactly because of (the danger of) this enforced introduction of ecological modernization into developing countries, industrialized countries in a globalizing world should abandon it.

Before dealing with these arguments and criticisms reflecting the Third World perspective, the focus of this critique of ecological modernization should be identified. These arguments are directed at the normative dimension of the Ecological Modernization Theory (or, at the reform models based on this theory) and leave the explanatory potential of the Ecological Modernization Theory for areas of the triad unchallenged.[17] This Third World criticism questions the adequacy of ecological modernization to present a viable program for managing the ecological crisis, given the conditions of unequal development in North and South, and the declining authority of nation-states to develop and implement their own ecological protection and reform model(s). However, it does not question the analytical-descriptive qualities of the Ecological Modernization Theory for those societies for which it was initially developed.[18]

On to the argument itself. These critics are of course right in noticing that in a globalizing world it is less and less easy to restrict certain ecological protection and reform strategies to a specific geographical area, let alone one country. When globalized "high-consequence risks" are involved and when environmental measures obstruct free trade or "fair" competitiveness, for instance, national environmental regimes are pushed toward convergence. Driving forces toward convergence might be international organizations and institutions, such as GATT and the WTO, the IPCC, the ITTO, and international pollution control arrangements. As a result of such convergence, the independent existence of fundamentally different environmental protection and reform designs in distinct nations or regions becomes increasingly anachronistic.

In spite of this, one can still question the validity and impact of the two arguments related to the Third World perspective. In balancing these arguments, I approach the central topic of this book. As to the first argument (the negligence of equal distribution of natural resources and environmental pollution), ecological modernization theorists claim that equal distribution of the "environmental (utilization) space" (Opschoor 1992) is by no means contradictory to their reform models. The fact that the present mode of environmental reform in the triad countries often disregards equal distribution does not imply that this aspect cannot be included in a radical ecological transformation of industrial societies along the lines of ecological modernization. Indeed, various Friends of the Earth reports from different Western European countries (Friends of the Earth Netherlands 1992; Spangenberg 1995) take the notion of equal distribution of "environmental space" or harmonization of environmental "backpacks" as the starting point for their models of environmental reform, based—explicitly or implicitly—on the notion of ecological modernization. These reports show that an ecological modernization strategy in industrialized societies is not inherently conditional on unequal withdrawals of natural resources or discharges of pollutants. Of course, the consequences of reform strategies that involve equal distribution of environmental resources are radical, both for existing patterns of production and consumption in triad countries (as the factor four/ten environmental programs show us; see, e.g., von Weizsacker et al. 1997) and for the political and economic international dependencies. In addition, ecological modernization strategies also revalue or upgrade resources to be used in international political and economics struggles: natural resources (such as metal ores, wood, and oil) and the environment (such as ozone layer, biodiversity and the greenhouse effect). These resources are utilized by non-triad countries to strengthen their position in international conflicts and negotiations, both in the field of environmental issues as well as in the fields of economics (chapter 5).

The second argument relates to what is rather overstatedly termed "neo-colonialism"[19] on the part of triad countries regarding environmental protection. This refers to the situation wherein the "traditional" structures and arrangements for environmental preservation in developing countries are destroyed or have become inadequate due to international interdependency,[20] and these countries are "forced" to import "modern" models of environmental reform from the triad countries that are ill-suited to existing local circumstances and the prevailing institutional order. However, it is doubtful whether the differences between the triad countries, and the (newly) industrializing countries in Central and Eastern Europe, Southeast and East Asia and Latin America will remain so large that the latter would be better off with fundamentally different environmental reform models and institutions than those employed in the triad. In this era of global modernity, with its high levels of time-space compression, the second argument can also be interpreted as supporting rather than undermining reform models based on the idea of ecological modernization.[21] Arguably, an "ecologization" of these countries in Central and Eastern Europe, East and Southeast Asia, and Latin America could take the slowly emerging ecological restructuring initiatives and models in triad countries as a starting point, no more and no less. To opt for a radical alternative outside the (globalizing) institutions of modernity may well mean throwing the baby out with the bath water. At the same time, European ecological modernization ideas and the corresponding environmental reform models are bound to take on a different form in non-European situations (Buttel 2000b; Frijns et al. 2000). In terms of the globalization literature: The homogenization of modern environmental reform will always go together with heterogenization, as general or global environmental reform dynamics become embedded in a specific way in local cultures, political systems and economies. The point I want to investigate is, to what extent globalization and harmonization induce variations of ecological modernization in (non-European and) non-triad countries.

Two conclusions can be drawn from this discussion. First, the strong objection made by what I have labeled Third World critics against the application of ideas of ecological modernization in both the industrialized triad countries and the NICs and CEE countries are not convincing. Second, this does not imply an unconditional embrace of the Eurocentrist Ecological Modernization Theory and the environmental reform models connected with it. Three arguments support a more cautious approach toward "ecological modernization" in settings outside the triad. First, as things stand today, these reform models seem inadequate for a number of developing countries. This is especially true for countries in sub-Saharan Africa and some of the least developed countries in Asia and Latin America. Their institutional orders differ substantially from those of the industrial societies of the triad, which means that major adaptations would have to be made before these environmental reform ideas, institutional designs and strategies are transferred successfully. Second, though falling under the umbrella of "ecological modernization," the environmental reform models and experiences of the triad countries are not identical, both in terms of institutional frameworks, long term strategies and practical experiences in different countries, and the matching between national and supra-national or global regimes (as will be seen throughout this book). To some extent, environmental reform will always carry the specific characteristics of the local institutional order, no matter how strongly globalizing processes interfere. Finally, under the present conditions of reflexive modernity, social practices and physical relationships with nature are continually being examined and reshaped in the light of new incoming information about those very practices and relationships. This means that no conclusive "answers" (or reform models) to environmental disruption can be given, either in developed societies or in non-triad countries. In a certain sense "muddling through" (Lindblom 1959) has indeed become the rule rather than the exception, albeit not at random but along rather narrowly defined trajectories that others and I have labeled "ecological modernization."

Globalization: Between Environmental Apocalypse and Ecological Modernization

These conclusions only make sense if globalization processes are in fact, either partly or largely, both the carriers of and triggers for environmental reforms. However, I began this chapter by noting that it was exactly the emergence, however marginal, of globalization in the environmental debate which has led many to deny the positive developments in environment-induced institutional reform, away from ecological modernization.

Should we thus conclude that ecological modernization ideas have limited value beyond the borders of a limited number of rich European or triad nation-states? Or are globalization studies and perspectives put up with a rather archaic set of environmental ideas and ideologies, which prevailed in the 1970s and the early 1980s, but which are no longer considered adequate at the present, late modern time? In investigating the dialectics of globalization and environmental reform, I do not intend to deny the value of an ecological modernization perspective in the global world order (a question rather to be investigated in this study than to be answered beforehand). Nor do I dispute at forehand the real environmental dangers and devastating consequences of contemporary (economic) globalization processes, so strongly linked to apocalyptic perspectives. What this study tries to do, basically, is to make sense of what at first sight would seem contradictory developments in the environmental debate, by acknowledging the value and the limitations of both. Emphasizing the detrimental environmental effects of globalization, and especially global capitalism, in chapter 4, and the global structures and actual global practices enabling environmental reform in chapter 5, I consciously harbor the dichotomy in order to clarify the contradictory perspectives. At the same time we have to acknowledge that in the majority of situations the most fruitful explanations are to be found somewhere along the continuum between the two extremes, albeit at different points for different social practices, localities, and times.

4

The Globalization of Side Effects

"Students of globalization must surely take seriously the possibility that underlying structures of the modern (now globalized) world order—capitalism, the state, industrialism, nationality, rationalism—as well as the orthodox discourses that sustain them, may be in important respects irreparably destructive." This quote from Scholte (1996, p. 55), illustrated by the ecological consequences of globalization, sets the pace for a critical evaluation of globalizing processes and tendencies that is in line with the dominant trend in the environmental debate. Most of the authors associated with this trend see the relationship between globalization and the environment as one of progressive environmental deterioration and diminishing scope for environmental management and reform. This chapter follows the logic of that perspective.

Environmental Consequences of Global Capitalism

The environmental consequences of globalization are usually related primarily to the globalization of economic production and consumption—in short, global capitalism. My analysis also begins at this logical point, although a more complete examination cannot be restricted to this dimension. The global outlook of chains of production and consumption, with economic relationships stretching over great distances and material resources and intermediary goods flowing through the world economy, increases the time and the space between the origins of environmental neglect and actual environmental consequences and deterioration in specific localities.

Various mechanisms are emphasized in identifying negative environmental side effects in an ongoing process of economic globalization. The most obvious relationship between globalization and environmental degradation follows from the general statement that economic globalization processes regarding trade, foreign direct investment, economic decision making, management concepts, financial markets, and so on go together with enlarged production and consumption and thus directly result in decreasing environmental quality. Although it is generally acknowledged that not all units of increase in production and consumption have the same environmental consequences, the idea of a general relationship between decreasing environmental quality and economic growth in an era of globalization is still widely accepted. As Runge (1994, p. 34) puts it: ". . . greater economic integration may lead to greater transportation needs, higher levels of manufacturing output, and general increases in the demand for raw and processed products, all of which impose greater wear and tear on natural ecosystems." For example, figures 4.1–4.3 clearly show the differences between various economic regions in regard to non-renewable energy use and the greenhouse effect. It is this line of argument in particular that is central to the work of scholars concentrating on unrestrained global competitiveness as the main element in globalization. Economic globalization can then be equated with neo-liberalism—that is, with almost unregulated free-market capitalism on a global scale, in which deregulation, privatization, and the absence of any serious "control" dominate economic processes. The Group of Lisbon (1995) in particular directs its social and environmental criticism against unconditional free-market competition on a global scale, but it avoids the mistake of equating this unrestrained free-market capitalism with globalization as such—a mistake severely criticized by, among others, Nederveen Pieterse. This criticism of unrestrained global capitalism runs somewhat parallel with the more conventional neo-Marxist-inspired criticism of unfettered capitalism that prevailed in the environmental debate in the 1970s. To be sure, such a line of argument makes sense and provides a valuable framework of analysis for contemporary processes of global deterioration in a number of cases. Let me introduce some examples of such links between environmental deterioration and the global capitalist order with its inadequate restrictions on production and consumption. The division of labor connected to processes of economic globalization results in an increase in transport of materials (natural resources, waste, capital goods, intermediary goods, and final consumer products) and in energy consumption and the emission of greenhouse gases (figures 4.1–4.3), and it aggravates the risks of major environmental

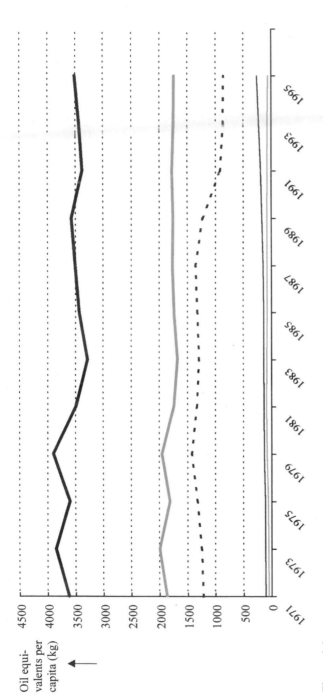

Figure 4.1
Consumption of non-renewable energy per capita (oil equivalents per capita, in kilograms), 1971–1995, for the United States (thicker solid line), the European OECD countries (thicker light line), Central and Eastern Europe (broken line), South-east Asia (thinner solid line), and West Africa (thinner light line). Source: http://www.rivm.nl/env/int/coredata/geas/index.html.

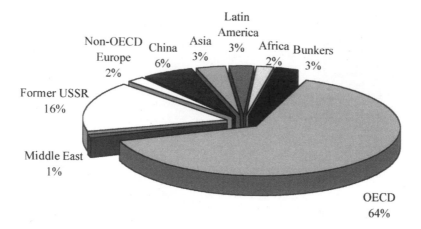

Figure 4.2
CO_2 emissions from fuel combustion, 1973 (total: 16,200 metric tons). Source: International Energy Agency 1998.

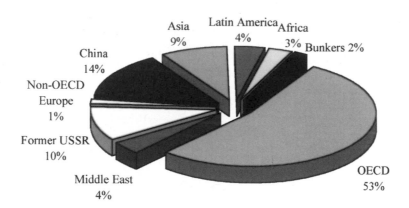

Figure 4.3
CO_2 emissions from fuel combustion, 1996 (total: 22,700 metric tons). Source: International Energy Agency 1998.

accidents. The same is true for the growth in movement of persons around the world, which is largely unaffected by the innovations of modern communication technology. This mobility not only results in growing emissions and natural resource depletion; it also causes local nuisance and claims on scarce space around airports. Economic globalization contributes also to loss of biodiversity by stimulating almost unconditional economic development in sensitive areas in developing countries, where most biodiversity is still located. In addition, the global diffusion and widespread introduction and use of machine technology, both for production and for daily consumption, will in a number of cases lead to aggravated exploitation of the environment due to, for example, increasing waste, emissions, and use of natural resources.

An example of empirical evidence of the growing environmental deterioration due to economic globalization is the study on dematerialization by Stephen Bunker (1996). Bunker criticizes the idea that continuing growth goes together with diminishing environmental impacts and resource use—part of the so-called environmental or green Kuznetz curve argument, which has recently become both highly popular and thoroughly despised. In his evaluation of this idea of dematerialization, Bunker looked at the volumes and distances of minerals transported over the period which the dematerialization thesis covers. He concluded that dematerialization was practiced long before the second and even before the first wave of environmental concern, and that the thesis as interpreted since the 1970s falls short of empirical data supporting the notion of general dematerialization of the global economy, is biased toward relative dematerialization (per unit GNP) or some local best practices instead of absolute dematerialization on a global scale, and does not include all local environmental effects related to extraction, processing, transport, final use, and disposal. Taking these latter effects into consideration would, according to Bunker, result in the conclusion that lighter product alternatives have larger proportional environmental impacts. According to Bunker, the globalizing world economy has not brought about a net physical dematerialization. Other writers, working from different empirical studies, have reached similar conclusions on the environmental consequences of an increasingly globalizing economy.

Of a distinct and more indirect nature are the environmental ills that go together with economic stagnation, crises, and depressions connected with the dynamics of the global economy. Referring particularly to the situation in sub-Saharan Africa in the 1980s, the Brundtland Commission has noted that uncontrolled economic stagnation and decline often result in increasing environmental deterioration. Both rural and urban environmental degradation in Africa are closely interlinked with the global economic processes of extraction. The 1997–98 economic crisis in Asia points to similar processes, albeit not of the same magnitude as in Africa. Though this economic crisis slowed manufacturing in Indonesia, the potential environmental gains of lower production levels were canceled out by increasing abatement costs and decreasing inspection and enforcement rates, with the net result of increasing industrial pollution (Afash 1998; Afash and Wheeler 1998). According to these authors, only major international environmental assistance and a shift from end-of-pipe environmental technologies and strategies to cleaner production can prevent further environmental deterioration in such situations of economic stagnation. The environmental consequences of the economic crisis in some of the transitional economies in Central and (especially) Eastern Europe and Asia point in the same direction (European Environment Agency 1998). Rising mortality rates in some of the former Soviet republics, reversing a worldwide historical trend toward increasing life expectancy, are believed to be partly caused by growing environmental ills.

Investments by Western companies in less developed regions can sometimes result in a better environmental record than the crisis-oriented production modes of domestic businesses. However, Western investments often stagnate in situations of economic crisis, as illustrated by the limited investment in sub-Saharan Africa by European and American TNCs. In addition, the environmental effects of foreign direct investment (FDI) are heavily debated and disputed in various arenas, under various topics and notions. Zarsky (1999) divides the FDI-environment linkages into three main categories:

• the micro-linkages, including the impacts of environmental standards on business location and investment decisions (also known as the "pollution haven" and "pollution halos" concepts) and the impacts of foreign business technology and management practices on firm-level environmental performance

• policy linkages, related to the effects of international competition for FDI on national environmental regulatory regimes

• macro linkages, including direct impacts of FDI on the scale of ecological degradation; impacts of FDI-related increases in income and consumption on ecosystems and environmental policy; impacts of FDI on local revenues and the provision of public goods such as treatment systems; socio-environmental impacts of FDI on workers, communities, and indigenous cultures; and environmental security impacts of FDI-related cross-border pollution and degradation on international conflicts and cooperation.

The net environmental benefit of FDI for the receiving nation is far from evident, the more critical analysts conclude.

These kinds of mechanisms of the "global diffusion of industrialism" have "created 'one world,' in a more negative and threatening sense than just mentioned—a world in which there are actual or potential changes of a harmful sort that affect everyone on the planet" (Giddens 1990, pp. 76–77). The near absence and inadequacy of mechanisms and institutions on a global scale that could correct or compensate for the excessive environmental consequences of global capitalism remind one of the national failures in environmental reform in the 1970s. Ecological modernization theorists have observed that—compared to the 1970s—such correction mechanisms and institutions are now better designed and function more smoothly at the level of the nation-state in industrialized countries. But their absence at the global level is still massive. This provides a partial explanation for both the relative popularity of neo-Marxist analyses of environmental deterioration at the global level and their marginal position in the national environmental debates in industrial societies. At the same time it may partly explain why the Ecological Modernization Theory is still believed to be more fruitful at the level of industrialized nation-states than at the global level.

These physical, material, or substance flow analyses of the environmental effects of economic globalization by an increasingly rapid circulation of materials, economic goods, and people should be complemented by studies of the increasingly global but still unequal (or heterogeneous) distribution of environmental consequences and of the decrease in possibilities for management and control of (global) environmental problems by the institutional arrangements of high modernity, most notably the nation-state and science and technology. Before I deal with these two points, I will examine why global environmental change prevails so strongly among the more apocalyptic perspectives on globalization.

The Apocalyptic View of Global Environmental Change

Most of the literature dealing with the environmental side effects of globalization focuses on those environmental threats that have recently been clustered under the heading of global environmental change or the somewhat different category of "high-consequence risks" (Giddens 1990). Various analyses provide evidence of the relationship between global environmental change and social globalization (Sklair 1994; Paterson 1996; Wilenius 1999). I argued in chapter 3 that it would distort our theoretical understanding and models if we were to limit the analysis of global change and environmental quality to this specific category or definition of environmental problems, which has dominated the environmental agenda for 10 years. The production, persistence, and distribution of "normal" environmental risks that prevailed over the environmental agenda in the 1970s and the 1980s, such as water pollution, solid waste (non-) disposal, local air emissions, or diminishing food quality, are also conditioned by globalization processes, albeit sometimes in a different way.

In spite of this conclusion, it helps if we understand why global environmental change is so often connected to globalization in a negative or apocalyptic way. I will mention six reasons for this, each on its own providing only a limited explanation but together presenting a clear understanding:

• These global problems emerged on the public and political agenda just as globalization ideas were coming up and being translated into political programs, in the late 1980s and the early 1990s. This contributed to a direct linking of the two.

• These environmental issues have a truly global outlook in the sense that both their causes and effects cannot be restricted to the territory of one nation-state. Most countries are connected to both causes and consequences in some way, albeit often not in a similar or equal way.

• Effective management of these global problems is often believed to be beyond the control of the "old" institutional arrangements of the nation-state, in contrast to the environmental problems dominating the public and political agendas in the 1970s and the early 1980s, so they are immediately linked to the phenomenon of globalization.

• These global environmental issues show the diminishing enlightenment character of science and technology. Science and technology have been discredited as instruments for understanding, and hence solving, the causal mechanisms that endanger the global ecosystems and human health.

- Global environmental threats are widely perceived as being triggered by processes of economic globalization in which, for instance, the international division of labor propels the flow of material goods and the capitalist competition diminishes private interest in achieving costly environmental reforms.

- Global environmental change or high-consequence risks reflect a heterogeneous and unequal distribution of risks and threats. On the one hand, this environmental risk profile deviates from the traditional economic profile of classes and countries, since it also threatens the well-off as environmental threats penetrate their secured economic bastions. On the other hand, the environmental risk profile increases the existing economic inequalities by adding ecological inequalities (as the marginalized peripheral regions and groups have insufficient resources to protect themselves against these threats: the "double risk" societies or groups—see Rinkevicius 2000). This means that both scholars arguing from a First World perspective as well as those giving prevalence to Third World interests emphasize the apocalyptic dimensions of global environmental change.

Hence, we should not be too surprised to find a very strong and apocalyptic link being drawn by both Western and Third World scholars between global environmental change and globalization. However, there is one major difference between these two groups of scholars. In contrast to the former, authors arguing from a Third World perspective often contest the one-sided emphasis on global environmental change by industrialized nations in international negotiations and politics, and in academic studies on global change. They believe other environmental categories which they consider more directly and immediately relevant for developing regions are neglected, such as water pollution, air pollution, and hazardous solid waste. For this group of scholars, globalization interferes in a similarly problematic way with both global environmental change and these "normal" environmental categories. It is the experiences, analyses and politics from triad countries that have created the distinction, which I have balanced in chapter 3.

Unequal Distribution of Environmental Risks

In a number of cases, not least of all involving "high-consequence risks," economic globalization means that environmental consequences are spread around the world in such a way that it becomes increasingly difficult to escape from them, no matter how rich or mobile one is. This is the case with

polluting substances that move fairly freely around the world in the ecological system (such as greenhouse gases, ozone depletion gases, nuclear
fallout, persistent pesticides and heavy metals in the oceans) or extractive
activities that endanger common future prospects (such as the loss of biodiversity). Global economics are always involved in such threats, and indeed
environmental risks are sometimes spread around the world even more
directly via economic mechanisms, as in the case of pesticides and additives
in international food chains, or toxic coloring agents in plastics and textiles.
However, does this mean that environmental risks are equally distributed
around the world among countries and classes?

The Risk Society Theory

On the basis of such examples as given above, the Risk Society Theory
points out the truly global character of today's environmental risks, and the
inherent apocalypse-blindness[1] of contemporary economic and political
institutions. According to Ulrich Beck—the founding father and still the
leading author of the Risk Society Theory—global environmental risks are
democratic in that they do not distinguish between classes, and traditional
class differences are no longer adequate for understanding the distribution
of these risks among the population. Who can escape the greenhouse effect,
mad cow disease, or the pesticides' "circle of poison"? Do not environmental risks often rebound on the producers of these risks?[2] And if the
distribution is unequal, are not the "class" divisions of environmental
inequalities then increasingly formed along new lines: the vegetarian against
the meat eater (Creutzfeldt-Jacobs Disease induced by mad cow disease), or
the outdoor worker against the indoor employee (skin cancer induced by
UV-b)? However, these arguments should be balanced. Buttel (2000a) and
others are right in criticizing Beck's overstatement of the dissolution of
classes in the distribution of environmental risks in late modernity: generally, the rich still have a better chance of protecting themselves against or
escaping from such environmental dangers, and not everybody or every
group or class is affected in the same way. Although it is true in some cases
that "wegreisen hilft letzlich ebensowenig wie Müsli essen" (in the end,
moving away helps as little as eating muesli) (Beck 1986, p. 97), locational
patterns, educational backgrounds, life styles, and economic protection
opportunities do make a difference when it comes to most environmental
risks. This difference is underlined by those authors arguing from a Third

World and environmental justice perspective on which I will elaborate below. However, the value of Beck's contribution to understanding the distributional consequences of environmental threats lies in his observation that socio-economic categories (classes) and environmental risks no longer run parallel by definition, as was often claimed in the neo-Marxist analyses of the 1970s. He is also noted for observing that all members of modern society have to somehow "deal with modern environmental risks" amidst growing uncertainty and the increasingly limited ability of the old institutions of science and politics to provide guidance on how we should live and behave. In conclusion, unequal distributions of environmental risks are still very relevant; however, these follow somewhat new distributional patterns in late-modern society, and these risks affect everyone in that all face uncertainties on how to cope with them. This is especially true in those cases where a growing complexity of interdependencies on an international or global level occurs, as for instance in cases of "high-consequence risks."

Unequal Distribution and Environmental Justice

Thus, the diversifying (or heterogenizing) effects of global economic mechanisms and processes also make themselves felt in the environmental field. However, the mechanisms by which the different environmental consequences and effects are distributed among groups, segments and geographical parts of the world are quite diverse. Goldblatt (1996, pp. 63–64) looks at the heterogenizing environmental consequences caused by variations in natural conditions and ecosystems. Homogeneous threats have diverging consequences due to different geographic and environmental circumstances. For instance, all small and flat islands are endangered by potential sea-level rise. However, the so-called small island development states and countries such as Bangladesh are particularly threatened by sea-level rise because, in addition to their natural conditions, the opportunities for protection against or escape from these environmental threats are unequally distributed, primarily along economic lines. And there exist more similar examples of unequal distribution of environmental risks along economic categories. Bunker (1996) claims that developing countries are often confronted by the dark side of dematerialization processes occurring in the core of the Western economies. They are saddled with the most polluting stages in the production line of a product. The dirty and material-intensive mining and natural resource extraction and transformation processes are located in the Third

World, while Western companies gain access to and control over natural resources via joint ventures, long term purchase contracts and various forms of subcontracting (the so-called new forms of international investment, NFI). In addition, these peripheral regions are confronted with declining terms of trade for their natural resources. Well known and widely quoted, but nevertheless relevant, are data showing unequal access to and consumption of natural and environmental resources (Redclift 1996; Yearly 1996; Redclift and Sage 1998; Held et al 1999, pp. 376–413). In addition, it is claimed that these inequalities have not only increased because of globalization processes, but are an inherent part of them. An example of a major mechanism in the economic globalization of manufacturing are the export processing zones which, according to Sassen (1994, p. 19) and Miller (1995, p. 145) not only provide favorable tax regimes, but also lenient workplace standards and the freedom to damage the environment without paying the cost. In relation to this it is also often argued (LeQuesne 1996, p. 67ff.) that, with the rapid growth of foreign direct investments, investments in environmentally harmful production processes and products are increasingly concentrated in developing countries and regions, where monitoring and enforcement of costly protective measures are of a low standard.[3] In conclusion, numerous authors agree that there is a geographically unequal distribution of environmental degradation, as not all parts of the world contribute equally to environmental problems, not all parts of the world are confronted by equal levels of environmental destruction and risks and not all parts of the world profit equally from the economic fruits reaped at the expense of environmental damage. This spatial distribution often runs parallel to economic categories; but not always.

Within nation-states the quest for equal distribution of environmental consequences along spatial, socio-economic and racial lines is often articulated in the call for environmental justice or environmental equality (Hurley 1995; Szasz and Meuser 1997; Ringquist 1997). Over the last decade economic class and race have proved to be essential features in risk distribution. This is reflected in the considerable number of American studies—closely related to active environmental justice movements—which have challenged Beck's thesis on the growing democracy of environmental problems. It is remarkable, as Szasz and Meuser (1997) conclude, that these national and sub-national American studies hardly ever refer to supra-national, international or global studies on the unequal distribution of environmental side effects.[4] On an international or global scale the equal distribution of "environmental space" (Spangenberg 1995), the "ecological footprint" (Wackernagel and Rees 1996), and the "ecological rucksack" (Schmidt-Bleek 1994), rather than environmental justice, are concepts employed to draw public and political attention to the correlation between economic poverty on the one hand and disproportionate environmental threats and unequal access to environmental resources on the other. It is precisely in this area of unequal global distribution of environmental goods and risks that Dependencia, neo Marxist, and World-System theories are celebrating a revaluation and revival. Whenever they focus on the environment, these theories have time and again emphasized the disproportionate negative environmental consequences for developing or underdeveloped regions, while these peripheral nations remain excluded from the fruits of global economic development. As regards industrial production, developing regions have relatively weak environmental monitoring and control measures; thus, products and technologies used in daily production and consumption can create greater ecological risks, as the Bhopal accident taught us.

Multinational or transnational corporations play an important role in the unequal distribution of environmental risks. As the motors of capitalist expansion transnational companies have never been given much credit in the environmental debate, and certainly not by neo-Marxist inspired traditions. The evolution of this environmental debate toward globalization and the concomitant uneven distribution of environmental risks has only confirmed the general view that transnational companies employ unsustainable practices, which have extended to and perhaps even persist mainly in peripheral regions. Some authors identify the trend toward deregulation of local markets and free trade as the main dynamics that enables transnational corporations to expand—or even concentrate—their unsustainable practices in developing countries (Daly and Cobb 1994). Others point to the centrality of the pursuit of profit and growth that drives them to operate in developing countries, while—still—preventing them from adopting sustainable production practices (Korten 1995; O'Connor 1998). However, it is also possible to identify a slow change in the debate on transnational companies and their environmental impacts in peripheral regions. Reviewing this debate, Moser and Miller (1997) conclude that, with the growing penetration of transnational companies into developing countries the argument has recently switched from the original question of *whether* there is any possibility that TNCs can advance sustainable development, to focus on *how* these companies might do so. This means in fact that the debate can be increasingly understood in terms of ecological modernization, having turned to issues such as in which sectors and regions transnational companies can contribute to sustainable investment patterns and how should they operate their businesses in an environmentally responsible manner.

Diminishing Scope for Environmental Management and Control

Processes of globalization also interfere with environmental quality in that they subvert the conventional institutional arrangements for environmental management and control which dominated the era of "simple" or high modernity. During that period, the politics of the nation-state and science and technology in particular played an essential role, alongside environmental NGOs, in both identifying and monitoring environmental change, and in reducing environmental risks. Some scholars, as I will illustrate below, claim that in the present phase of late or reflexive modernity these conventional institutional arrangements are increasingly being eroded and becoming less and less effective in dealing with (new) environmental problems under the new institutional conditions. In addition, these authors assert that the new international or global institutions being established to take over these environmental management and reform tasks are inadequate for various reasons. These assessments contribute to the idea and perspective that globalization saddles us with an increasing number of unequally distributed and unmanageable environmental risks.

State Failure Revisited

In the 1970s and the 1980s many sociologists and political scientists took a critical view of the ability of the "capitalist state" to intervene in economic processes of production and consumption. The specific qualities of the "capitalist state" restricted both the type of interventions and the effectiveness of these interventions (Habermas 1973; Jessop 1984), including those involving environmental deterioration and management (Jänicke 1986). New ideas and analyses on globalization arising in the 1990s strongly reinforced this line of argument and Wapner can be seen as a recent representative of this position. Wapner (1995, p. 48ff.) divides the arguments regarding state failure into four main points:

- On a national level, the state fails to protect the environment in a systematic and proactive way.[5]
- States have signed many international agreements but compliance is poor and violations dominate.
- Even if these treaties are implemented they are inadequate since they often represent the lowest common denominator for environmental reform.
- The effectiveness of international agreements is doubtful because monitoring and verification are carried out by the states themselves.

The emergence of various supra-national and global political and economic institutions and mechanisms undermines state sovereignty and further reduces the capacity of states to intervene in those economic activities of production and consumption that degrade the nation-state's territorial environment. What is more, the limits imposed on management and control are not only related to those global environmental problems that transcend the national level, such as the greenhouse effect, the ozone-layer depletion and loss of biodiversity. The declining autonomy on, for instance, import restrictions and export bans (of hazardous waste), and the introduction of economic incentives to stimulate ecologically sound products also affect the state's ability to manage "normal" environmental problems. This is partly due to restrictions put on the nation-state's degrees of freedom by international institutions and organizations, such as those involved in global trade (GATT and the WTO) and those related to regional economic regimes (such as the EU and NAFTA). However, it is also caused by global economic dynamics, where states can no longer afford to ignore capitalist competition, since it determines most of the global allocation of capital. Summarizing this argument, Clark (1996) refers to a group of scholars who interpret international environmental politics as a "race to the bottom." This is a situation whereby nation-states are caught in a kind of prisoner's dilemma, as they strive to create favorable conditions in order to attract international capital. Global competition will then force international standards to group around the lowest common denominator, and individual countries will be motivated to move even below that.[6] Authors who are less skeptical, and more aware of the limited importance of national environmental standards and practices as investment factors, analyze the effects of international capitalist competition on national environmental regimes not so much in terms of a "race to the bottom," but rather as "stuck in the mud." This is a process whereby a "race to the environmental top" is inhibited by the fact that states are reluctant to take unilateral leaps toward better environmental management and more stringent environmental standards (Zarsky 1997, 1999). However, it is not only the growing international competition and the downward pressure on environmental standardization and harmonization that frustrates the strengthening of conventional environmental regimes of the nation-state and infringes on the environmental sovereignty of states. National environmental regimes also fail due to the growing organizational and technical complexity of the globalizing production and consumption systems; the high speed and wide-ranging character of social transformation that is difficult to regulate by rather inert, static, or slowly changing national environmental regimes; the fast flows of new information on the causes and effects of environmental problems; and the growing demand for the refinement and specification of general environmental protection regimes according to local physical, social, or economic conditions. As Lash and Urry (1994) extensively argue in their analysis of the changing relationship between nature and society in the era of globalization, contemporary nation-states are too small for the large problems, such as "high-consequence risks," and too large for the small problems, which require local fine tuning and close collaboration with local actors.

In looking for alternative political models that can deal more successfully with environmental threats in an era of globalization, various ideas have been presented, two of which merit further attention. Some scholars (e.g., those quoted in Frankel 1987 and Wapner 1995) argue in favor of the development of sub-statist solutions or "subpolitics" (Beck 1994a). Subpolitics are forms of non-institutional and sub-systemic politics that exist next to and often below (that is: more decentral to) the traditional national political institutions of parliament, bureaucracy and political parties. Held (1995) and the Commission on Global Governance (1995) are inclined toward cosmopolitan or global governance as an alternative solution. The two sets of ideas have at least three common factors: (1) They originate from existing initiatives, practices and experiments that present alternatives for the failing state-centered environmental governance models. (2) They all still have a long way to go before being put widely into practice and really begin challenging the dominant system of national environmental governance.[7] (3) All of them have been met with criticism regarding their feasibility, their effectiveness, and their desirability. The two alternative political models will be elaborated in more detail in chapter 5; in the remainder of this chapter, I will pay particular attention to the critics of the two alternatives.

Sub-Statist Institutions as Solution?

Sub-statist answers to the shortcomings and failures in environmental management and control by nation-states hinge on bringing together the various parties and interests involved in environmental breakdown at the sub-national level. The basic idea behind sub-statism is that the structures and institutions that alienate people from their natural environment as well as from their fellow citizens can only be overcome by downscaling the interdependent relationships to sub-national levels. The majority of sub-statist ideas have their roots at least partly in the pre-globalization era, and it seems that globalizing tendencies have not made them much stronger. Largely in line with earlier analyses, such as Frankel's (1987), Wapner (1995) identifies five objections to sub-statism:

• If these solutions work they will take considerable time to become effective, and it is doubtful whether sufficient time is available.
• Sub-statist models will create opportunities for small areas of the land or territory, but it remains unclear who will be responsible for the relationship between the parts, the whole, and the entire planet.
• Human activity is becoming more global rather than more local in scope, which sheds doubt on the adequacy of local solutions.
• It seems highly unlikely that nation-states will relinquish their authority to sub-statist organizations and/or authorities.
• Sub-statist ideas and models assume that people are essentially cooperative and ecologically mindful but that global (capitalist) structures and institutions alienate them. (This assumption is questionable according to Wapner.)

The concept of sub-politics is of rather recent origin and its emergence is often attributed to the failures of the traditional political (and scientific) institutions in the face of globalization. We have argued before (Hogenboom et al. 2000) that state involvement—especially but not only in the field of the environment—is criticized and bypassed by environmental sub-politics outside the traditional institutions in those cases where governmental environmental policies remain restricted to natural science definitions of environmental problems and risks, leaving questions of public perceptions, norms and values unaddressed, are immobilized by internal conflicts of interest and legitimization or by stagnating international engagements, remain restricted

to measures facilitating negotiation or coordination between conflicting societal interests without attempting to solve them, and/or prove to be too
bureaucratic, rigid and universal to meet the complex diversity and local
connotations of environmental problems. A rich literature reporting on
experiences regarding solutions to environmental controversies by agencies
outside the state has recently emerged, with examples coming from several
countries (Bosso 1998; Nownes 1991; Rawcliffe 1998; Stafford and
Hartman 1998; Murphy and Bendell 1997; Hogenboom et al. 2000;
Duyvendak et al. 1999; Mol 2000), but the first significant experiments with
environmental sub-politics in countries such as the Netherlands date back to
at least the early 1990s. These sub-political arrangements and ideas have
been criticized from different angles. Some scholars interpret sub-politics as
the last resort of institutions of modernity in their attempts to deal with their
own detrimental side effects. These authors disapprove of sub-politics for
not making a radical break with modernity. While celebrating the analysis
of an emerging Risk Society, Bauman (1993; 1997) vehemently opposes the
idea of sub-politics as a solution to the detrimental effects of the global modern order. Sub-politics, he believes, will only strengthen those institutional
traits that are inextricably bound up with the Risk Society, and will thus
have a counterproductive effect in the end. A second group of authors asserts
that environmental problems triggered by globalization processes would be
difficult to deal with via sub-political processes, since sub-politics is—or was
originally—mainly located at the sub-national level. This comes close to the
sub-statism criticism quoted above. The creator of the notion of sub-politics,
Ulrich Beck (1996), has partly responded to that criticism by elaborating a
concept of "world sub-politics," but neither his conceptual framework, nor
the empirical evidence he presents[8] of sub-politics at the global level are very
substantial. Yet others criticize sub-political solutions for not being democratic (because the conventional representative institutions are bypassed),
transparent, or effective (Duyvendak et al. 1999).

Supra-Statist Solutions under Fire
The second alternative, the search for cosmopolitan or global governance
as a possible way of managing environmental problems in an era of globalization, is also received with reservations. The school of thought represented by authors like Wolfgang Sachs (1993) is rather critical of the
possibilities of global environmental management and governance. All

attempts to come up with global initiatives to manage environmental disruption, whether initiated and advanced by states, transnational corporations, international institutions or non-governmental organizations, are doomed to have further negative side effects—be they ecological, social or gender-related. Their solutions usually point in sub-statist directions. So when Shiva (1993) makes a plea for the democratization of global institutions such as the World Bank, she insists that "the 'global' must accede to the local, since the local exists with nature, while the 'global' exists only in offices of the World Bank/IMF and the headquarters of multinational corporations." For these scholars the alternatives often lie in a withdrawal from global institutions, into the local, into the South, into resistance. Although some of this criticism on global environmental governance and management is also reflected in a number of recent studies on the future of the Bretton Woods institutions (Cavanagh, Wysham, and Arruda 1994; Griesgraber and Gunter 1995; Griesgraber and Gunter 1996), their final conclusions and options for institutional and environmental reform are not always so pessimistic and "anti-global." The main critical arguments of this group are that global environmental governance by these institutions has failed because the institutions likely to support them are too weak (most UN institutions), whereas the stronger institutions, such as IMF, WTO and—to a lesser extent—the World Bank, are not very concerned about environmental goals. The WTO, seen by McMichael (1996b, p. 169) as the epitome of globalization, has global governing powers based on an abstract set of free trade rules, an unelected bureaucracy, confidentiality and zero citizen participation. Both democracy and the environment are jeopardized by such forms of institutionalized economic globalization, and it should therefore not surprise us that the fiercest censure of these institutions comes from environmental NGOs, labor unions, and other social movements. However, in the debate on proposals for future reform of both the powerful and the powerless global organizations, the environmental consensus splits into three groups: those looking for the institutional reform of the stronger (often economic) institutions along environmental and social lines, those who perceive this as merely strengthening the powerful institutions at the expense of other international (UN) agencies more open to public environmental pressure, and those who promote bottom-up, local, or national solutions without addressing these global agencies, which they perceive as irretrievably lost for the cause of sustainability. One of the more detailed and well-balanced ideas in the latter school of thought is Martin and Schumann's (1996) plea for re-strengthening the democratic nation-state and restoring the primacy of national politics over economy.[9] Drawing on numerous examples—and touching incidentally on environment-inspired arguments—they show the devastating consequences of globalization processes for the non-elite, in particular, in the North and the South. The only way out of this globalization trap, they feel, is the enforcement of democratic state structures, which are mainly—but not necessarily only— restricted to the national level. As for democratic state structures above the level of the nation-state, they make a plea for a strong and democratic European Union that can counterbalance the economic globalization processes in world capitalism. These arguments are similar to the stand-points and ideas of the Group of Lisbon (1995) and, if less closely, also to those of David Held (1996). I will come back to these views on global governance in chapter 5.

Science and Technology

Science and the epistemic communities are sometimes put forward as a new global authority that could trigger political institutions into successful action for (global) environmental reform. This perspective stems from the idea that the nation-state and conventional political institutions dealing with the environmental crisis have been undermined, and that international or global governance institutions are nonexistent, unsustainable or politically marginal. However, the view taken in this chapter suggests that science and technology themselves have been discredited as institutions for the management and control of (global) environmental problems. With global environmental change and the greenhouse effect as examples, various authors within the Risk Society school of thought have drawn attention to the fact that science has been unable to present any lasting conclusions as to the consequences and solutions of increasing greenhouse gas emissions into the atmosphere. Science and the epistemic community have for many years behaved and been seen as an independent authority in questions of environmental deterioration (i.e. from the 1960s till at least the mid 1980s). Under the present conditions of reflexive modernity, however, science has lost its enlightenment character. It is not just that scientific research is no longer able to produce irrefutable evidence on causes and solutions, as is often illustrated citing the example of the greenhouse effect and the scientific work of the Intergovernmental Panel on Climate Change (IPCC). Scientific expertise as such, and the role of science and scientists in the further development of industrial society, are now being questioned by lay actors and politicians alike. Inadequate handling of uncertainties and lack of trust in scientific risk and safety calculations often prove to be the rule rather than the exception in complex global, and sometimes synergistic interactions. The "mad cow disease" is just such a case where scientific evidence was inadequate in restoring confidence about meat consumption among politicians and consumers. Genetic modification of food is believed to be the next candidate in a growing row. In addition, according to Beck and others, science and technology are heavily involved in the imminent endangering of modern society by producing, preserving and even legitimizing environmental risks, while the attempts to manage these threats by means of science and technology actually increase ecotechnocracy. This is the case with, for example the IPCC global circulation models and their inbuilt forms of politics (Beck 1996, p. 6). As a consequence, science and technology (or "expert systems" as Giddens calls them) have become the object of anxiety-induced suspicion and ontological insecurity in the public at large.

Against this background the original idea that—in the absence of adequate political systems at the level of the nation-state or else forceful global governance—science, institutionalized in the epistemic communities, could evolve into a kind of supra-national authority that might "force" nation-states and other actors in the global arena to take common action on global environmental problems, is being abandoned, both on theoretical grounds and on the basis of empirical evidence. In fact, critics claim that the emancipating role of science and technology in identifying, defining and setting the agenda with regard to environmental deterioration, or in formulating precautionary and preventive solutions has been increasingly discredited.

Intermezzo: Environmental Side Effects and the Compression of the World

In conclusion, the notion that relatively unrestrained global competitiveness and economic growth contribute to increasing environmental deterioration is not contested. Economic globalization has led to an aggravation of negative environmental side effects. Empirical studies on the material dimension of the contemporary globalizing economy indicate that the overall global trend is in fact not dematerialization and more ecologically rational use of environmentally sensitive goods, but rather the opposite. This may not be true for all sectors, countries or regions. However, the overall pattern is growing environmental deterioration, unequally caused by and distributed among the different nations of the world. In this message lies the value of neo-Marxist, World System, and anti-globalization studies.

Increasingly, however, the focus and debate in globalization-and-environment studies are moving away from the inherently detrimental environmental consequences of global capitalism, toward other issues. To what extent is this current global capitalism really unrestrained (or neo-liberal), and to what extent should it be? How many and what kind of (environmentally relevant) correction mechanisms and constraints can and should be put on global capitalist development now and in the future? In addition, what are the possible environmental and socio-economic consequences and gains of proposals and ideas for more restrained and "conditional" global economic development? The fact that these questions are now being asked marks a shift from a position against any idea of globalization, toward a perspective that acknowledges that globalization is happening but distinguishes different forms and outlooks.

A large variety of proposals have emerged for combating the detrimental effects of (particular forms and aspects of) globalization on environmental quality and strengthening environmental management. The most radical proposals and solutions build upon the demodernization, deindustrialization and sub-statist theories of the 1970s. These argue for shortening the chains of production and consumption in order to limit, for example, transport and thus energy consumption, and the risk of major accidents, for raising the standards of product quality and control and, thus, restoring confidence in these products among consumers (Sachs 1993). However, these solutions to the problems created or aggravated by globalization fail to take account of the actual and dominant forces that carry globalization processes forward. Without being deterministic, one can still acknowledge that these globalization forces cannot simply be brushed aside or reversed should we so desire.

Paradoxically, some of the forces that contribute to further globalization are in fact environmental threats, actual and perceived. Ecological threats are not simply caused, prolonged, aggravated, or mismanaged due to globalization processes. Via various economic, cultural and political mechanisms, the environmental threats also call forth processes of globalization in various areas, and thus add to the compressing of the world. As Steven Yearley (1996, p. 28) states, "The world's growing environmental problems are connecting the lives of people in very different societies." David Held (1995, pp. 105–107) goes even further, arguing that concern for our common heritage undermines the traditional Westphalian model of international relations. Environmental problems, both global and local, have definitely become important cornerstones of global culture and consciousness, and have in that way triggered and/or accelerated global political and economic processes. Environmental movements have not only sprung up around the world, they have also become globalized in the sense that parts of the environmental movement are trying to set a joint global agenda for action, and that truly global organizations, such as Greenpeace, the World Wildlife Fund, and Friends of the Earth, have emerged. Arguing along these lines, Strassoldo (1992) identifies the ecological world view as one of the central factors in the processes of globalization. In accordance with this view, I will turn to the other side of the relationship between globalization and the environment in the next chapter.

5

Globalization and Environmental Reform

Lash and Urry (1994, p. 3) are right when they suggest that "the sorts of economies of signs and space that became pervasive in the wake of organized capitalism do not just lead to increasing meaninglessness, homogenization, abstraction, anomie and the destruction of the subject"—nor, I would add, only to environmental destruction and failing environmental management. A considerable number of authors within the field of globalization theory emphasize—to some extent with valid arguments, as I have pointed out in the previous chapter—the disastrous effects of globalization processes on the environment and environmental management. However, there is also another, contrasting side. The institutional transformations of modernity in the age of globalization are not all detrimental to the environment. The perspective or theory of ecological modernization can serve to focus our analysis on those mechanisms and institutional transformations of modernity that contribute to environmental reform, and supersede the idea of an all-pervasive juggernaut of globalization whose negative side effects are inevitable and beyond all "control."

New Logics of Environmental Reform

As I have discussed in chapter 3, the development and growing popularity of the concept of ecological modernization from the mid 1980s on, reflects actual transformations in environmental debate, practices and relevant institutions in a number of OECD countries. Environmental concerns and environmental reforms have moved beyond the initial stage of window dressing and marginal corrective measures in the periphery of the environmental crisis in industrial societies. They are now approaching the core institutions that are widely held responsible for the environmental threats to modern industrial society. Up until now, almost all analyzes of these environment-induced changes have been restricted to the level of the nation-state, or at best to a number of closely related industrial nation-states. However, if we are to take the globalization debate seriously, the nation-state can no longer be the only, or even the principal, unit under study. Global flows of information, capital, ideas and people, worldwide economic, political and cultural networks and the development of new institutional arrangements at the supra-national and global levels occupy a central position in globalization discourses and contrast the primacy of the nation-state in sociological analyses. Analyzing the synchronicity of globalization and ecological reform in the perspective of ecological modernization, the essential innovation is then not so much the mere upscaling of this perspective from a national level—for which it was originally developed and to which it has been most often applied. Rather, the creative challenge is to apply a perspective that was originally developed in order to study industrial nation-states to an investigation of processes of ecological reform connected with globalization processes and dynamics. Hence, it is not just a matter of scale but also a matter of quality. Chapter 3 makes it clear that this difference hinges on the changing character of modernity. In a era of reflexive modernization, as opposed to high modernity, ecological transformation processes follow a different "logic," which can only be fully understood by moving our attention beyond the practices and institutions of the nation-state.

When considering the synchronization of globalization and environmental reform two sides of this coin should be borne in mind. First, and further to the argument that concluded the former chapter, ecological concerns, motives and interests can be a driving force behind processes of globalization (in the economic, cultural and political domains). The concern for the preservation of the environment will, in a number of cases, reinforce globalization. Similarly, and secondly, globalization processes can strengthen environmental reform and environmental consciousness. Globalization can trigger the harmonization of national environmental practices, regimes, and standards, produce new institutional arrangements at a supra-national level, transfer environmental technologies, management concepts, and organizational models, and accelerate the exchange of environmental information around the world. This chapter focuses on this kind of synchronization or mutual reinforcement between globalization and environmental management and reform. Consideration is given respectively to new environmental arrangements related to the globalization of economic manufacturing, the modernization of political institutions and practices beyond the nation-state, and the emergence of global environmental norms and values, movements and consciousness.

Global Ecological Restructuring

Analyses of global capitalism often highlight the unrestrained nature of the activities that are believed to have such harmful effects on the environment in various parts of the world. However, in the daily practice of worldwide economic institutions, systems and relationships, this very same global capitalism shows another dimension, in which a diversity of economic and non-economic mechanisms jointly promote or contribute to global ecological restructuring. In order to underpin this latter perspective, I will concentrate on transnational enterprises and global economic institutions as these are two important global economic entities that are not only involved in the deterioration of the environment, but also foster innovations in institutional systems that protect the environment.

Transnational Enterprises and Environmental Reform
The changing relationship between nation-states and transnational enterprises, and especially the power shift from the former to the latter in the global arena, is often regarded as one of the causes of severe environmental deterioration, both at the national and the supra-national or global level. Murphy (1994, p. 57) rightly criticizes the naive position of Brown (1972) and others unconditionally celebrating the positive role of transnational companies in global environmental reform, claiming that these enterprises choose the most efficient combinations of productive resources. Thirty or more years of in-depth theoretical and empirical analysis of markets and their main actors have made it clear that efficient resource allocation is a theoretical construct that hardly ever reflects real-life situations and carries little relevance for environmental preservation in view of the present system of exchange values of resources. But though criticism of such unconditional faith in neo-liberal market capitalism and its main actors (TNCs) is justified, there is another side to the dynamics of transnational enterprise in global capitalism.[1] In recent years in particular, transnational enterprises,

the most important actors in the globalization of manufacturing, have frequently been driving forces behind the harmonization of national environmental regimes (see below). In addition, they have pushed for global environmental regimes and are progressive implementing agencies for, among others, environmental management and audit systems (EMAS), new environmental technologies, cleaner production methods, and new organizational principles that reflect environmental concerns. Their powerful position as spiders in economic webs allows them to trigger environmental improvements and reform in client and supplier companies. Despite the fact that a number of global conglomerates still show contrasting environmental practices in their various local branches (e.g., Union Carbide in Bhopal and the United States, Shell in Nigeria and the Netherlands, Nike in Vietnam and the United States), transnational companies show a tendency toward a harmonization of environmental practices. This convergence is encouraged by such things as international standards, global information networks, international regulations and global liability, codes of conduct and civil pressure. It is exactly their global dimension that make these enterprises vulnerable from an environmental point of view, not only on the sites where environmental mismanagement takes place, but in all corners of the world economy where they manufacture goods and make their products available. The Shell Brent Spar case—in which the dumping of an oil producing platform in the Atlantic was prevented by consumer boycotts in numerous countries—has made this especially clear. Nike, too, has experienced the problems of diverging standards in different locations (O'Rourke et al. 1996; O'Rourke 2000). A situation has arisen in which "the stateless firm has an interest in transnational forms of authoritative decision-making which reduce regulatory divergence between nation-states, yet the existence of the stateless firm gives a further impetus to the extension of the regulatory role of transnational political structures" (Grant 1993, pp. 72–73).[2] Global enterprises have a clear interest in a homogenization in environmental reform and therefore promote it. In practice, contrary to what the more economically oriented rational choice theories have often predicted, this does not often result in a "race to the bottom"—a homogenization in the environmental performance of global firms at the lowest common denominator (Clark 1995). Global capitalism cannot be equated with the "prisoner's dilemma" situation of a free market in which global companies operate without any regulation, communication or norms. Environmental safeguards beyond the "bottom line" have been promoted by regulations, communications and norms both inside and outside the global economic domain.

The example of international standardization puts this dynamic into perspective and prevents us from being overly optimistic. The International Organization for Standardization's 14000 series and the British Standard 7750 are examples of regional or global harmonization of environmental practices in manufacturing. Falke (1996) shows how these standards can contribute to regional harmonization by shifting environmental standardization practices from the national to the European Union level. Analyzing the emergence of global standards in environmental performance via the ISO series, Roberts (1998a) concludes that these standards trigger environmental improvements in products and processes in companies in the triad, as well as in the larger exporting companies and their suppliers in newly industrializing economies and developing countries. National industries and the informal sector in the latter group of countries have appeared unaffected by these global environmental standards up until now. In a detailed analysis Riva and Gleckman (1998) warn us not to expect too much from ISO 14001 (the standard for environmental management systems): "Firms, regulators and the public are far from sure that ISO 14001 on its own will bring environmental performance improvements or sustainable industrial development."

Economic homogenization and standardization are essential to the success of global "green" firms, such as The Body Shop, and they have also promoted the greening of multinationals in developing countries and regions (Robins and Roberts 1997; Weinberg 1998). Other well-known environmental initiatives are several global and voluntary business programs, the participants in which aim to be—or let others believe them to be—responsible producers in respect of the environment: the Responsible Care initiative of the chemical industry, the Business Charter for Sustainable Development under the aegis of the International Chamber of Commerce, and the Coalition for Environmentally Responsible Economies (CERES) which has adopted ten principles for environmental management (Adams 1999). Scholars such as Karliner (1997) and activists who portray these initiatives as an exercise in public relations or strategic communication have a point. Most of the initiatives and developments in this direction certainly have such elements, and we cannot fully understand the emergence and functioning of individual "green" initiatives, companies or organizations, such as the Business Council for Sustainable Development, if we do not take media considerations and public relations into account. However, strategic communication and public relations can no longer remain merely an exercise in image building. They must correspond to actual environmental improvement and performance in order to survive and avoid damaging the reputation—and sometimes market position—of "green" organizations.

As a rule, transnational enterprises cannot be ranked among the top polluters, not in the developed countries and even less in developing countries, as I will explain in more detail in chapters 7 and 8. They are usually too visible, vulnerable and "standardized" to be prepared to risk environmental confrontation in one locality that may rebound on their global sales and production. This depends to some extent on the sector and the "profile" of the firm. According to Moser and Miller (1997), high-profile extractive industries often working in pristine environments, and firms in the (petro)chemical sector face the strongest pressure to operate in an ecologically proper way. Similarly, high-profile consumer goods producers are often highly conspicuous on the issue of labor conditions. Zarsky (1999) adds that large foreign firms with close links to green or environmentally sensitive consumer markets perform better in environmental terms. In contrast, the larger nationally oriented firms, primarily confronted and sometimes even protected by national regimes, are more likely to be among the massive polluters, especially in non-triad countries. Environmental reform of these national industries has taken on an aspect of hybridization rather than homogenization, as global norms and standards have mixed with varying national and local policies and interpretations, and with specific company cultures. Stringent domestic or international environmental norms are often seen as a hindrance to the competitiveness of national companies, both by their managements and the national political elite. In contrast to that, Vogel (1997) argues that theoretically stricter domestic regulatory standards can actually give domestic businesses a competitive advantage, as they can more easily comply with them than can foreign businesses. Others have labeled that phenomenon regulatory competition.

The Greening of Global Economic Institutions
Within an ecological modernization perspective we can also interpret and understand the increasing advance of environmental considerations by some of the global and regional economic organizations that have tradi-

tionally advocated the global free market. These include the OECD, the Economic Commission for Europe, the European Union, NAFTA, and, more recently, the World Bank (Kolk 1999). This process of environmental institutionalization started from an economic logic, responding to the demands for a harmonization of environmental requirements and standards in the various nation-states as a means to reduce unequal competition. However, environmental policy making in these institutions has gained its own momentum and internal "logic," which can no longer be reduced to a narrow economic rationality. This is evidenced by the environmental agreement added to the NAFTA treaty; the environmental performance reviews and suggestions contained in Expanded Producer Responsibility ideas (Lindqvist and Lifset 1998) launched by the OECD; the growing volume of environmental EU directives; the World Bank's inspection teams and the cessation of its support for large dam projects; the emergence of "environmental conditionalities" in development aid and debt relief (McAfee 1999); and the rights of citizens to become engaged in or complain about some of these institutions. Some scholars have analyzed these developments within economic institutions in terms of the emergence of (environmental) proto-states. Although these international economic institutions have indeed appropriated some of the activities that nation-states initiated on the eve of the second wave of environmental concern, the differences are obvious and the emphasis should certainly be strongly on the *proto* in proto-states. In addition, the environmental reforms of these economic institutions should not be seen as evolutionary developments unfolding smoothly in this new millennium. The "greening" of these international organizations is accompanied by constant struggles and pressures, both from inside and outside these economic institutions. The Multilateral Agreement on Investment (MAI), as proposed by and negotiated within the framework of the OECD from 1996 to 1998, provides evidence of the force of countervailing economic powers that can sometimes, as in the case of MAI, only be overcome by a major effort and pressure from forces outside the economic domains (see also chapter 7 below).

Some global economic institutions such as the World Trade Organization (the successor to GATT) and the International Monetary Fund—by no means the insignificant ones—are hardly making significant progress in combining the advancement of global trade, investment, and economic growth (on one hand) with stringent environmental norms and reforms (on the other). In these institutions, economic "free" market capitalism still dominates as the undisputed and undivided rationale. The attempt to include stronger environmental provisions in the last Uruguay round of GATT negotiations failed, although it did pave the way for the next round of negotiations. Nevertheless, proposals for environment-informed restrictions on imports and trade are meeting less and less principal opposition in this forum, and the discussion seems to be moving toward issues of proportionality and the unjustified use of environmental arguments in order to protect national economic interests. These two institutions increasingly face difficulties to ignore the growing external pressure to include environmental considerations in their activities and decisions. It has become common understanding—even within the WTO—that global economic institutions are involved in global environmental legislation and policy making, whether they want to be or not, and the fact that environmental interests and liberalization cannot be fully separated is therefore being acknowledged. Thus, despite critical comments on the environmental records of the IMF, the World Bank, and the WTO, Goldenman (1999) is probably right in claiming that the International Financial Institutions (IFIs) and the multilateral banks are being forced to increase their transparency and their accountability for environmental side effects, and that they are changing their role in global governance.[3] Increasing demands for "economic justice" by broad coalitions of civil society groups and movements are putting these economic institutions under pressure. According to Goldenman, the national export credit agencies—whose role in catalyzing private sector finance is even greater than that of banks and IFIs—operate without any transparency or accountability and are rarely subject to scrutiny by powers that might redirect their activities along more environmentally sound paths.[4]

In conclusion: Environmental reform through global economic institutions can no longer be understood or explained by economic reasoning only. The environment is slowly becoming a relatively independent factor in these new institutional arrangements, albeit not equally in all of them. The strategic question in regard to global environmental reform remains whether such institutional reforms of existing economic organizations should be the main goal (as advocated by the "greener" member states of these economic institutions) or whether strong global environmental "counterpart" organizations or institutions with a sufficiently strong mandate, power, and influence should be established, as was proposed by Esty (1994). Both alternatives can contribute to homogenization in global environmental reform.

Political Modernization beyond the Nation-State

One of the most obvious contributions of globalization processes to the strengthening of environmental reform is to be found in the harmoniza tion or "homogenization" between national environmental regimes in lesser developed regions and those in triad countries. Meyer (1987, p. 48) underlines that what he calls isomorphism is a striking characteristic at the present time: "Peripheral societies shift to modern forms of industrial and service activity; to modern state organizations; to modern educational systems; to modern welfare and military systems; in short, to all the institutional apparatus of modern social organisation." Although this suggest too much an evolutionary convergence, we can indeed see in most "peripheral" and semi-peripheral states the construction of political institutions for environmental reform that are clearly inspired by the experiences and institutional examples of triad states. In the late 1980s and the early 1990s, most developing countries began to develop environmental legislation and environmental planning and establish environmental agencies and authorities at the national and sub-national levels. The global dynamics of the third wave of environmental concern, not least the 1992 UNCED conference and its followup, gave a tremendous boost to that development. However, numerous drawbacks have been pointed out by scholars. It is often emphasized that in "peripheral" regions these institutions rarely function properly; that they have become necessary only because the path of industrialization was forced upon them; that their construction has been to a large extent promoted by and beneficial to Western and international consultancies, donor agencies, IFIs and Western environmental technology industries; and that these institutions have caused a weakening of environmental systems in triad nation-states by reducing harmonization to the level of the lowest common denominator. However, the net effect—especially in the so-called newly industrializing countries—is most likely to be positive, certainly in the long term.

These national political developments in "peripheral" states are not the only environmental innovations occurring parallel to globalization. Globalization and the declining autonomy of the nation-state—interpreted by some as disastrous for environmental reform—are accompanied by new modes of governance and politics, also—or perhaps especially—in the environmental field.⁵ The "new politics of pollution" (Weale 1992) are partly being triggered by processes of globalization. Globalization trends have reduced the effectiveness of the traditional national political institutions dealing with environmental reform, and as a result these are being transformed and modernized into new institutional arrangements. The reframing and redefinition of relationships among political institutions, the market, and civil society at a global level has led to the emergence of new political arrangements at the national and the subnational level and at the supranational level. Although such new institutional systems have emerged in various arenas, I will concentrate on those related to environmental reform.

(Sub-) National Political Innovations

At the national and the subnational level, two developments in environmental politics can be distinguished, both related to globalization: the transformation of national environmental governance styles and the decentralization and localization of environmental governance and reform.

The first innovation entails the reform of traditional and hierarchical command-and-control environmental regulation or governance styles, which are increasingly perceived as being inadequate and ineffective in successfully controlling the environmental crisis. Factors that contribute to the undermining of the effectiveness of traditional environmental institutions include new circumstances of international competition, increasing liberalization in the circulation of goods, the greater complexity of producer-consumer relations stretching beyond national borders, a highly dynamic and rapidly changing social context of environmental policy making, the exponential increase in new information on the causes and consequences of environmental deterioration, and a growing awareness of the ambivalence of that information. These "globalization" factors have strongly contributed to the situation whereby negotiation, cooperation, interaction and consultation between parties at an early stage of environmental policy making are now more often the rule than the exception in the triad countries. Such tendencies of so-called political modernization can be witnessed in all triad countries, though these have often been adapted to the prevailing national policy styles and cultures. Notions of environmental mediation, negotiated rule making, regulatory negotiations, joint environmental policy making, voluntary agreements and communicative regulation all apply to the phenomena of political modernization.[6] Even in developing countries similar political innovations in environmental management have emerged, sometimes exported by leading industrialized nations such as the Netherlands (with all the teething troubles that go with such transfers, since the specific local conditions and institutional layout have often not been sufficiently taken into account), and sometimes introduced by transnational companies that bring an experience of these policy innovations from the industrialized world (Jones 1997; Doyle 1999).

Some scholars interpret these innovations in environmental governance as forms of "sub-politics" (as introduced in chapter 4). However, these innovations differ from "sub-politics" in that they center on national or state environmental policy. "Sub-politics" involves politics that move beyond state politics and its traditional political institutions, but it will always remain linked to them (see below). Critics of sub-politics are right in asserting that it can never fully replace the "old" conventional politics of the nation state. Still, sub-political arrangements are on the advance in environmental reform in industrialized countries, since they remedy some of the drawbacks of the environmental politics of nation-state institutions under conditions of globalization. The question should be asked whether these, partly globalization-induced, political innovations in the field of the environment can be considered improvements, or merely "logical" or inevitable consequences that will eventually undermine effective environmental reform. There is no single answer to this as it depends heavily on the specific constellation and distribution of power around these innovations of national political arrangements. Comparing between political innovations in the Netherlands, Denmark, and Austria, we found clear differences, ranging from deregulation-inspired ideas to major improvements in environmental governance along the lines of "political modernization" (Mol, Lauber, and Liefferink 2000).

A second, partly connected, sub-national development related to changing global state-market relations can be characterized as decentralization and local diversification. In the era of globalization many regions and communities are faced with evaporating national protection mechanisms and changing institutional systems due to such developments as (neo-liberal) deregulation, privatization and decentralization. Diversification in production, consumption and policy making has become a "survival" strategy to handle or even escape the destructive tendencies of globalization. This than has to include a reorientation on local conditions, both natural and social, and more freedom of maneuver for local producers and politicians. Numerous authors (especially in the field of rural sociology and food studies) have provided various examples and experiences of diversification in agriculture and food production. Most of these studies celebrate the beneficial ecological—and social—consequences of such local diversification strategies and decry the devastating ecological consequences of global food systems, despite compliance with detailed (national and supra-national) regulations, monitoring, inspections and quality control.[7] The mad cow disease, the dioxin poisoning of Belgian chicken meat and the deliberate hormone additions in American beef are cited as critical examples of the failures of global food systems. Apart from the question of whether such an interpretation and the corresponding preference for localization is always— or even in most cases—valid from an environmental point of view,[8] local diversification can only be interpreted and understood fully in connection with globalization processes, no matter how much local practitioners or their academic advocates wish to separate them. It is ironic that French farmers opposing these global food systems because of its health risks and lack of quality see their markets threatened by high American import taxes on typical French products, such as local cheese varieties.

Supra-National Political Innovations

Globalization involves the emergence of new supra-national political institutions and arrangements regarding environmental issues, both at the regional and the global level.[9] In short, globalization means the end of the primacy of national political arrangements. Political modernization in the era of globalization can never be reduced to national or sub-national political modernization. Economic globalization of the processes of production and consumption and the undermining of the nation-state's autonomy or sovereignty go together with the transfer of capacities for environmental policy making and governance from the nation-state to international or global institutions, the increasing importance of non-government actors (such as transnational enterprises and environmental NGOs) on the international stage of environmental politics, and an increase in the formation of global or regional environmental regimes, which in turn trigger movements toward homogenization. The nation-state is still an important actor in national policies on, for instance, solid waste, surface water pollution control, nature reserve protection, and soil pollution cleanup, and nation-states will remain crucial actors in international environmental politics for some time.[10] However, the modernization of environmental politics is moving in the direction of political globalization. In this process, the failure to modernize national environmental policies may rebound on the nation-state via global institutions. Germany's reluctance to experiment with emission trading was overruled by the Kyoto agreements on Climate Change, in which the EU pushed for experiments with emission trading both within countries and globally. Germany might well be "forced" to join in emission trading via its EU membership.

The establishment of supra-national environmental regimes is of course advancing particularly rapidly in the European Union. In these times of serious criticism of the EU and calls for an intergovernmental rather than a more federal model, the unique European case of supra-national politics is often defended with an appeal to environmental arguments. This defense claims that a strong European Union is not only essential from an economic point of view, but especially necessary to protect the environment in a unifying market where environmental problems transcend the level of the nation-states. Indeed, in the EU of today, environmental politics do surpass the authority of the nation-state, as discussed in chapter 6, although states still have a major say—but diminishing powers of sovereign veto. This model of supra-national environmental politics *in status nacendi* is often put forward as the model for constructing new global institutional arrangements on environmental reform. To be sure, EU experiences are unique and valuable in promoting the further development, refinement and transformation of the present state of fragmented global environmental regimes toward more consistent and integrated global political systems that restrain global capitalism. After all, that is the bottom line of what most scholars aim for: taming environmental problems related to global capitalism via global institutions and systems that are better equipped to do this than the present weak UN system. Several authors have contributed ideas to the design of a future system of "global institutions for sustainable development" (Commission on Global Governance 1995; Esty 1994 on trade and environment). Some start from the current situation and refine it on a case-by-case basis (Dorfman 1991, and the numerous contributions to the development and refinement of more focused, issue-specific environmental regimes[11]). Others work toward EU-like regional institutions which have sufficient power and mandate to counter the devastating environmental effects of regional and global competition (Group of Lisbon 1995). Finally there are those who take a more "creative" and long-term perspective as their point of departure. Harris (1992) proposes a radical restructuring of global institutions to meet the future needs of sustainability, since strengthening the authority and capacity of current environmental agencies will not be sufficient to tame global capitalism.[12] No matter how desired, proposals such as those of Harris (1992) suffer from some of the weaknesses and limitations that—according to Nederveen Pieterse (1997)—are often found in constructive approaches to global regulatory and institutional reform: It is not clear how these proposals relate to ongoing transformations. They often reflect weaknesses in institutional analysis. They often pay little attention to the relationship between these global reforms and reforms at local and national levels. They do not satisfactorily combine diverse interests and perspectives that could promote new institutional structures.

There is a second innovation in supra-national environmental politics, especially related to the non-triad countries. The fact that states, especially the poorer ones, increasingly use environmental resources (both in terms of natural resources to be exploited and in terms of emission "rights") in the world struggle for economic resources and power (Miller 1995), is both a consequence of the increasing importance of international environmental institutions and a reflection of the necessity for more powerful global environmental arrangements.[13] With the establishment of environmental regimes at the global level, the environment has become a powerful resource for nation-states, not only in terms of compensation for the burdens imposed by stringent international environmental regimes, but also because it can help them secure additional economic and technological rewards. As Sachs et al. (1998) argue, without the less developed countries "international agreements on protection of the climate cannot be achieved, and they can threaten global repercussions if they follow the Northern example and industrialize without concern for the consequences and rob the earth of its sinks as well." The debt-for-nature or debt-for-environment exchanges, joint implementation, the clean development mechanism and other flexible implementation measures—put forcefully on the global agenda by the Northern countries during the 1997 Summit on Climate Change in Kyoto, Japan—are examples of the use of environmental resources by non-OECD countries for securing both economic and technological gains as well as national environmental conservation. Porter and Brown (1991, p. 134) rightly note that at both the 1972 Stockholm and the 1992 Rio de Janeiro summits, developing countries secured financial revenues and technology transfer as a kind of compensation for environmental agreements and measures. Recently, similar "compensations" have been negotiated in global arrangements on CFCs, forests, climate change, and biodiversity. In that sense, developing states have something extra to gain from the formation of global environmental politics and regimes.

Global Civil Society: Movements and Discourses

Chapter 4 ended with the observation that environmental deterioration, and the social dynamics linked to it, have triggered a general awareness of global interconnectedness. It is global environmental risks, in particular, that have played a role in transforming existing environmental consciousness among certain sectors of the population in the triad countries into three directions: It has evolved into global (environmental) awareness. It has spread among much larger segments of the population in triad countries. It has increasingly spread to large segments of non-triad countries (although "normal" environmental risks played an equal important role in that latter development). Of course, the awareness of global interdependence did in fact emerge much earlier among environmentalists (in the 1970s), when it was popularly articulated as "think globally, act locally" and "from one earth to one world." The emergence of global environmental change on the public and political agendas not only caused this awareness to spread out among broader segments of industrialized societies, but also to infect major parts of the Third World elites and middle classes. Next to global environmental change and other environmental threats it is the worldwide appearance of the ecological world view that most authors of globalization theories emphasize when they analyze globalization in relation to the environment. "Globalisation has helped to increase ecological consciousness and programmes to enhance sustainability," Scholte (1996, p. 53) writes, before embarking on a lengthy discussion of the global environmental threats related to globalization. However, the reverse is also true, since ecological consciousness has contributed to globalization trends. Global polls (such as the worldwide Gallup polls on the environment), the growth and growing influence of global environmental NGOs such as Greenpeace, Friends of the Earth and the World Wildlife Fund, and the rapid rise of more local environmental NGOs in industrializing and developing countries are all believed to demonstrate the globalization of ecological concern. Most authors who look for a global green countervailing power against the destructive effects of unfettered global capitalism, point to the emergence of what has become known as a global civil society (Commission on Global Governance 1995; Lipschutz 1996; Wapner 1997).

A Global Environmental Movement

Outlining the political struggles on the environment at the level of the nation-state, some authors either identify or advocate similar initiatives on a global scale. International environmental NGOs[14] with their global campaigns, lobbying, demonstrations and boycotts are seen as the global actors who can translate the widespread environmental awareness and concern into international political pressure (Princen and Finger 1994; McCormick 1995, 1999). These ideas are included in works such as Arts's (1998) study on the political influence of global NGOs on the Biodiversity and Climate conventions following the UNCED process, and Wapner's (1996) study on the role of Greenpeace, the World Wildlife Fund, and Friends of the Earth in what he calls world civic politics. These studies rely heavily on similar investigations of national environmental movements, in that they look at the number of organizations, their composition in terms of origin, class, income and so on, the main resources used in their struggles, as well as their strategies and ideologies. This implies that the mechanisms by which environmental concern was channeled into organizations and, hence, political pressure at the national level are seen as models for developments at the global level. It goes without saying that, although such a comparative analysis might offer us insights into the emergence of this global counterpower, the dynamics of an emerging global civil society are and will be very different from its national predecessors. This is not only or primarily because of the differences in scale. For one thing, there is no global authority equivalent to the nation-state, and nation-states will remain crucial in the global arena. Also, there is much greater heterogeneity at the global level, both as concerns the various segments of civil society (e.g., regional and national varieties) and among those who are addressed by civil society (e.g., states and polluters). Provided these differences are taken into account, as Wapner

(1996, p. 160ff.) does, such studies on global environmental movements can provide us with insights into global civil society's contribution to environmental reform.

Discourse Coalitions and Universal Values

Two highly interesting theoretical concepts have been developed in studies on the mutual relationship between globalization and what might be called the environmental discourse. Maarten Hajer (1995) has used the concept of discourse coalitions in his analysis of the changing national environmental discourse in some triad countries. Discourse coalitions are the configuration of a set of story lines, the actors who articulate these story lines and the practices in which the discourse is rooted and reproduced.[15] The essential element in discourse coalitions is that dispersed groups with varying ideas and belief systems on the environment gather together around a specific discourse, a common political project. Usually the dominant story line that is the mortar in a discourse coalition reduces the discursive complexity, even though it necessarily has a multi-interpretable character. Sustainable development, or ecological modernization, can be interpreted as such a concept on which the dominant discourse coalition on the environment centers nowadays.[16] Interpretations and definitions of environmental problems, solutions to these problems, and the strategies to be developed converge more and more around these dominant concepts and related story lines. According to Weale (1992) and Hajer (1995, p. 14), the processes of globalization and the perception of a global order in particular have promoted a common environmental story line among various "partners" of the environmental discourse coalition.

The second concept for analyzing the relationship between globalization processes and environmental reform is the development of universal norms and principles as proposed by Steven Yearley (1996). One of the major issues, and also a major environmental achievement that runs parallel to globalization processes, is the development and institutionalization of universal principles in global environmental discourses, practices, and political strategies. Such universal principles may be founded on intersubjective standards and/or scientific knowledge. However, it is widely acknowledged among scholars that, with the disenchantment of science in this era of reflexive modernization, universal principles based solely or primarily on scientific knowledge have increasingly become untenable. The 1992

UNCED and the initiatives following it are often perceived as parts of a process in which environmental norms and principles have acquired inter-subjective agreement. Ideas launched by the UNCED increasingly reappear in all kinds of international agreements, treaties and standards, resulting in the development of more consistent international environmental legislation and regimes. The Precautionary Principle is an example of a concept that is increasingly gaining global recognition, albeit not without debate and struggle. Its successful acceptance within the Biosafety Protocol (agreed upon in 2000) of the Convention on Biological Diversity is in stark con-trast to the failure of attempts to get it included in the WTO provisions (Cameron 1999). A second example is the declaration of the Århus Conference in the context of the Environment for Europe process. The European (both EU and other) ministers for the environment gathering together in Århus, Denmark, in 1998, agreed to accept three core principles as universal guidelines for environmental governance: access to environ-mental information, public participation in environmental decision mak-ing and access to judicial remedy (Petkova and Veil 2000). One of the attractions of such universal norms and principles is that they appear to be expressions of worldwide authority and provide some sort of "solid" foun-dation on which to deal with conflicting interests at the supra-national or global level. This is particularly important in circumstances where global government is still far away and science is losing rather than gaining ground in its attempts to become an alternative global authority (chapter 4).

Interpreting globalization processes with the help of these two concepts, it can be seen that a basic—though perhaps still low-level—common inter-subjective understanding has emerged regarding the necessity of protecting environmental values, and the way to do it. A convergence of environmen-tal discourse coalitions and universal norms and principles are essential ele-ments of the growing influence of ecological rationality in global decision making and global developments. In fact, however, so-called universal envi-ronmental discourses that are believed to be emerging are not so universal, since they generally encompass diverging sectoral, national, or regional or class interests, as became clear during the 1992 UNCED and the environ-mental discourse on sustainable development.[17] Similarly, the global envi-ronment polls that are believed to express increasing agreement on the priority given to safeguarding high environmental quality might measure quite different perceptions and definitions of the environment in different locations. The "universal" ecological world view can be expected to differ and diverge in different local and cultural contexts. Nevertheless, although there is much truth in these more balanced views on the converging tendencies of global civil society as regards the environment, two major accomplishments have been achieved since the green slogan "think globally, act locally" became popular in the late 1960s. First, global awareness and debate on environmental deterioration and reform have expanded from small, local, and peripheral environmental groups in the triad to large and important sections of world society. This expansion has been triggered by, among other things, the inability of the conventional institutions of simple modernity (e.g., the nation-state, science and technology, the liberalized market) to control major categories of environmental risk. Second, activities by green action groups have also become globalized, albeit not without obstacles (Rucht 1993). Global civil society and its environmental vanguard organizations (environmental NGOs) have a major role to play in establishing, designing, influencing, and altering global political institutions, as Wapner (1997) shows. By the same token the shortcomings of the conventional institutions of simple modernity have contributed to the emergence of global "sub-politics": new institutional arrangements on a global level that bypass traditional political institutions by linking diffuse environmental consciousness within civil society with global decision making in the economic and political arenas. Some go a step further and relate global civil society to the emergence of "global governance without government," where the institutions that have made up national and international environmental politics for so long are absent. Examples of such innovations are fair and green trade pushed by civil society organizations, private global enterprises that collaborate with development NGOs, liberalized utility markets in search of both renewable resources and green consumers, and incidents such as the Brent Spar. However, these developments have remained quite limited and marginal in scope until recently.

Toward "Global" Ecological Modernization

Each perspective—the globalization of side effects and the globalization of environmental reform—has analytical value, as I tried to show in the previous two chapters. Both perspectives offer us insights into the two environmental sides of the same globalization coin. The central question is, of course, which side is dominant in what kind of contemporary regional and global social practices and supra-national institutional developments. As I emphasized in chapter 1, most contributions by academics and environmental activists to the globalization-and-environment debate are quite straightforward in emphasizing the devastating environmental consequences of globalization processes, while taking a skeptical view of environmental reform, either actual or potential.

In the attempt to identify, understand, and contribute to the global processes and dynamics of ecological modernization, I do not wish to contradict the perception and the position I outlined in chapter 4. The logic of the arguments and the empirical evidence cannot be denied with regard to the present international order. However, I do think that this view needs to be balanced, since contradictory developments have occurred in the contemporary world order, although perhaps less strong in force and impact as some might hope or believe; and that building on the current developments of ecological reform is a promising and perhaps the only alternative to counteract the devastating environmental consequences of today's economic globalization. Therefore, in the following chapters, I will take a closer look at the practicalities and possibilities of strengthening ecological modernization mechanisms and processes in social practices and institutions that are heavily "governed" by globalization. Starting from the assumptions that globalization processes are a reality and that ecological reform can be advanced under conditions of globalization, it is rewarding to make an attempt to help reorganize and re-design the process of globalization into more ecologically sound directions. Meanwhile, however, this book does not aim to outline a political program or a concrete design for new organizations or institutions at the global level (Young 1997a). By analyzing and emphasizing existing practices of "global" ecological modernization—frequently in *status nascendi*—I will rather show the current possibilities (and limitations) of strengthening these dynamics.

6

Globalization as Triadization: The EU, NAFTA, and Japan

Most of the economists working on globalization acknowledge that the process of globalization does not have the same meaning and the same economic consequences throughout the world. Scholars as diverse as Dicken (1992; 1998), Ruigrok and van Tulder (1994), the Group of Lisbon (1995), and Castells (1996; 1997a,b) focus on the concentration of the movement of financial capital and foreign direct investments in a limited part of the world: the so-called triad of the European Union, the North American Free Trade Agreement region, and Japan and some of its neighbor "tiger" economies in East and Southeast Asia. It is especially in these developed regions that the characteristics associated with economic globalization emerge. Political and social scientists as well as economic scholars draw the same conclusion: (economic) globalization is not actually global—or planetary, as Castells (1996, p. 92) calls it—in character. Some scholars, however, such as Hirst and Thompson, push this conclusion one step too far, claiming that we might have to look for another, more adequate concept.

It is to these core regions of globalization that I now turn. If these "triad" regions are in the forefront of economic processes of globalization, a closer analysis of the interaction between such economic processes and dynamics of environmental deterioration and reform in this triad will substantiate the theoretical considerations of the previous two chapters. The fact that economic globalization is concentrated in the triad does not mean, of course, that its influence and side effects—especially regarding the environment—remain limited to these three developed parts of the world. On the contrary, many scholars have not only pointed out how more "peripheral" economies are linked to globalization in the triad, they also emphasize the unequal distribution of the environmental consequences of globalization. They have shown both that the most well-off regions manage to export or relocate the negative side effects to less well-off areas and population groups and that global environmental policies are clearly designed to benefit the well-developed world, while the poorer parts bear the negative (economic) consequences. I will concentrate on the (environmental) interplay between the triad and the developing economies outside it in the following chapter. In this chapter the "internal" global dynamics of the triad will take a central place.

Although the starting point will be economic globalization—as in most globalization studies—it goes without saying that the political and socio-cultural dimensions of globalization have to be included in such analyses and empirical illustrations. Although some extreme neo-liberal economic scholars as well as some radical critics of the present world order still want us to believe otherwise, we are not living in an unregulated global free-market capitalism. Political actors, widespread norms and values, and other more reflexive processes are constantly active—with ambivalent results—to redirect global economic processes into less harmful directions. If we are to understand the "real life" consequences of globalization—in contrast to the theoretical possibilities following from some kind of economic or other model—it is essential to take those non-economic dimensions fully into account, beyond the level of seeing them as incidental successful or unsuccessful attempts to change the dominant course of global capitalism that will never be able to succeed in any structural way.

How has each of the three sides of the triad dealt with the ongoing pressure on the natural environment following the changing character and patterns of economic globalization or global capitalism? As was shown in chapter 2, all three regions of the triad (and the countries within these regions) are experiencing similar economic developments, and are increasingly interdependent in terms of global competition (that is competition as regards the cost of labor, the investment climate, regulatory policies, physical and non-physical infrastructure, etc.). But they also show differences in such things as their political culture and institutions, their socio-cultural traditions and civil society, their natural environments and their degree of economic and political integration. The similarities and interdependencies are of course the foundation for the neo-liberal demands for deregulation, liberalization and decentralization, and on the other hand for the neo-Marxist and otherwise inspired denunciations of the detrimental external effects of globalization. Together these similarities and interdependencies are believed to be culminating in tendencies toward homogenization. The differences are associated with heterogenization and counterreactions (both at the level of the nation-states and beyond) against economic globalization in those regions and arenas where homogenization and its side effects are perceived as dominant and pervasive. I will analyze each of the three areas, with special attention for these contrasting developments. The three regions also differ in the sense that the EU and the NAFTA regions have internal economic—and increasingly environmental—harmonization and integration policies and governance, while the Asian side of the triad shows hardly any supra-national economic or other integration policies or governance. Consequently, the analysis is not so much performed at the level of the nation-state but primarily—and especially with respect to the EU and NAFTA—beyond that.

Supranational Homogenization: The European Union

Before elaborating on the environmental response to economic globalization within the EU region, we have to be aware, first and foremost, that the EU should be interpreted primarily as an institutional design to enlarge the internal market, with increasingly permeable borders and a harmonization of economic regulations. If anything, the original purpose of the EU was to improve the region's economic functioning (it was the European *Economic* Community) by harmonizing economic policies and instruments and lifting trade barriers, and to formulate an answer to both the economic successes of the new powers (the United States and Japan) and the internal economic and political conflicts. Most of the legislative and policy-making efforts of the European Union have always been directed at the opening of the Common Market and the harmonization of the nation-states' economic policies. This was believed to boost economic growth, improve competitiveness, enhance economies of scale, provide beneficial conditions for "domestic" producers, and so forth. Regional integration was always in the first place economic integration, driven primarily by economic logic and only secondarily by political and cultural motives.[1] More recently, non-economic integration processes have moved up on the agenda of the EU, sometimes as a logical consequence of economic integration and sometimes owing to pressure by non-economic advocacy coalitions (e.g., coalitions addressing the social agenda, worker protection, cultural issues, the environment). Only after 1970 did environmental policy making begin to appear in the development process of the EU, without any constitutional basis at first, and firmly anchored in the EU's treaty only since 1987. Although originally the environmental activities by EU institutions were primarily related to the economic harmonization of regulatory regimes in the member states in order to ensure equal competition and a level playing field, such activities were soon interpreted more broadly and less contingent on an economic logic, as is shown by the emergence of for instance directives on bird protection and the water quality of bathing resorts. In the pursuit of common environmental policy, the connection with economic harmonization became looser and environmental protection became a common goal, sanctioned by the Treaty revisions that have become known as the Single European Act (coming into force in 1987), the new European Treaty of Maastricht (coming into force in 1993) and its revision of Amsterdam (1997). This later development is to some extent under pressure following the emergence of the subsidiarity principle,[2] though that principle threatens to affect all major common policy areas and not only the environment.

The European Union is often cited as being unique in its supra-national qualities. It is the only international organization that has the power and the legal right to limit the sovereignty of its member states. Although the European Union will not easily become a European sovereign power, i.e. a kind of federal state, it has moved well beyond an association of sovereign states. From various perspectives—and not lastly from an environmental point of view—the European Union is often cited as exemplary for the future design of international organizations that have to confront global problems (Held 1995; Martin and Schumann 1997; Beck 1997). The European region is therefore of major interest for this study for two reasons: it was the first geographical region where the free transnational movement of goods, capital and people was greatly stimulated and *de facto* accelerated (processes that are now often associated with economic globalization), and it is the first region were an environmental answer to these accelerated international economic trends has been formulated in terms of supra-national environmental politics. With respect to socio-cultural transnational processes, on the other hand, the EU seems to move "forward" rather slowly: identity formation, transnational or European new social movements, European political parties and the formulation of European norms and values are processes which seem to lag behind Europeanization in the economic and political dimensions.

Economic Liberalization and Environmental Risks

Now two sets of—interrelated—questions seem especially relevant in assessing the relation between "globalization tendencies" and environmental deterioration and reform. The first focuses on the contribution of inter-European developments to the aggravation of environmental risks. Do processes of opening up the borders, increasing trade, harmonizing economic and political regimes, and accommodating national environmental policies to economic competition, add to environmental risks? The second set of questions relates to the kind of mechanisms (or achievements) that counteract or regulate economic globalization dynamics. It can be national as well as supra-national, political as well as socio-cultural or economic mechanisms. I will deal with these two sets of questions respectively.

The European integration process is still to a major extent an economic integration process, designed to create a common market without barriers and stimulate the economic development of the member states, and especially its producers. Although there are too many interfering variables for scientific proof of causality, the common understanding is that the European integration process has indeed contributed to economic development in the region, making the region more competitive toward the United States and Japan. It is not by accident or some mistaken understanding of self-interest that major European producers are usually among the first to argue for a next step in the economic integration process (e.g., a common monetary system). Parallel to—but not necessarily due to—the European integration process and the wider process of globalization, most European economies have shown continuous economic growth resulting in an increase in transportation (persons as well as freight), industrial and agricultural production, energy use and consumption. This increase in economic activities is accompanied by a process of concentration of production, processes of privatization and liberalization, and increasing foreign direct investments. In fact, what we see are the typical characteristics of globalization patterns, be it on a smaller, regional, scale. As often, there is no one-to-one relationship between the growth in economic activities on the one hand and environmental deterioration and risks on the other. The European Environment Agency (Stanners and Bourdeau 1995; European Environmental Agency 1998) shows that increasing transport or production are not necessarily followed by similar increases in pollutant emissions to air or water, or an equivalent decrease of environmental quality, due to technological changes, political interventions, economic concentration, organizational transformations, and the like. In some cases, we witness not only a delinking of economic growth from environmental parameters, but even an absolute decline of the latter. More likely than not, we will never be able to conclusively answer the question: will the economic integration of Europe result in further environmental devastation? Nevertheless, this discussion has accompanied the Europeanization process for some time.

The discussion whether the European Union is ultimately a blessing for the environment or the Trojan horse that will only prove to have sped up unrestrained economic growth and liberalization, and impaired efforts to control economic investments and production, re-emerges from time to time, and varies considerably from country to country. Opponents of the European integration process as well as parts of the environmental movement have strongly questioned the consequences of continuing economic integration for the environment, particularly at the crucial junctures in the formation process of the European Union, such as the negotiations on the Single European Act or the revisions of the treaty in 1992 and 1995.[3] In the environmental front-runner countries or pioneers (Denmark, Germany, the Netherlands, and more recently the Nordic member states and Austria; see Andersen and Liefferink 1997) the discussion was most vivid. Denmark's original refusal to accept the Maastricht Treaty was to a major extent inspired by environmental concerns, and most of the Danish comments on the Treaty of Amsterdam concentrated on the environmental consequences of this new step in the integration process. Environmental movements in these EU environmental front runner states have originally been equally skeptical of the integration process, and many of them have voiced anti-European integration standpoints at the various crucial moments in the integration process. More recently, however, a slow turn is to be observed in the environmental debate: most "mainstream" environmental movements no longer oppose the European integration process but rather press for a "greening" of this process. There are several reasons underlying this shift. First, Europeanization is well on its way and although there is still debate about the final target or goal of this integration process, it seems a "passed station" to just oppose any integration. Second, the nature of the integration process has changed and widened considerably. In various smaller steps, the environment has developed from a merely peripheral policy field in the 1960s and the 1970s toward a more central focus of attention that can no longer be ignored in new rounds of economic integration. Thirdly, the environmentalists themselves have changed. They have become more involved in Brussels politics and have adopted a more European perspective, noticing that while certain developments might look as a step backwards or standstill from a domestic point of view, they can be considered a major environmental step forward in a European perspective. The environmentalists from "laggard" states such as the United Kingdom or the Mediterranean countries in particular have gained an experience of the possibilities of using the EU to force national governments into environmental action. Finally, environmentalists have begun to acknowledge that while the EU is far from perfect from an environmental point of view (or a democratic point of view, for that matter), it is nevertheless the best option available at the moment in the way of supra-national institutions that might be able to control or reduce the environmental side effects of economic globalization. The evident need for supra-national political institutions is what makes the new "celebration" of the EU understandable (Group of Lisbon 1995). Both the leading German "green" politician, Joska Fisher, and the leading German environmental sociologist, Ulrich Beck, converge in their analysis that while the European Union was initially and is still primarily built around an economic integration project, it remains the most promising supranational political institution to challenge neo-liberal globalization tendencies, both within and beyond Europe: "Ohne Europa gibt es keine Antwort auf Globalisierung" ("Without Europe there is no answer to globalization") (Beck 1997, p. 261).[4]

EU Answers to Neo-Liberal Globalization

The consequences of all this are that more recently discussions on the EU and the environment focus less on the desirability of the EU integration project as such, but more on the actual design and the possibilities, instruments and mechanisms to turn the (still primarily economic) integration project into greener directions. This brings us to the second set of questions regarding the mechanisms and achievements that can be identified in counteracting the devastating consequences of economic globalization in the region. It should be noted here, that the political EU institutions engaged in designing environmental policy and reform are considered to be highly influential and exceptional compared to the institutions in the other two triad regions. But although EU environmental institutions and reforms have been relatively "successful," most environmental commentators remain rather skeptical in their assessment of EU environmental policies.

First, most environmental commentators, both from the EU and elsewhere, consider EU environmental policy to be rather fragmentary and scattered, failing to come up with a more systematic and overall outline of an EU environmental reform program that can seriously counteract the often devastating influence of the dynamic process of economic integration. The environmental action programs—the fifth (launched in 1993) being the latest—give the impression of some coherence, vision and direction, but they should be judged as rather toothless exercises, compared to the EU measures that really make a difference: Directives and Regulations.[5] The development of these latter environmental instruments seems to follow a logic that is definitely not related to any environmental basis or systematic plan. Second, most commentators rightly emphasize the difficulty of implementing EU policies and the limited powers of EU institutions that can only monitor and enforce formal implementation of EU directives in national legislation, instead of having a say on the actual implementation in terms of environmental improvements and standards in the field.[6] Third, commentators from the environmental pioneer states, as well as representatives from environmental NGOs, often emphasize the poor environmental ambitions of the EU and the "race to the bottom" argument: sometimes environmental EU directives are much less stringent than the measures preferred by or functioning in these pioneer states, even forcing these to lower their standards and goals to the level of the lowest common denominator. But this seems to be an overly narrow or one-sided interpretation, since the rather undeveloped or lax environmental regimes in for instance the Mediterranean states are just as often pushed upwards by EU directives (La Spina and Sciortino 1993).[7] And, in addition, several strategies have been developed (especially after 1987) to deal with the front runners'—but also the laggards'—predicament: differentiation in environmental speed (usually limited in time), minimum harmonization or the permission to go further than the EU, exceptions, and a redirection toward more general and procedural directives to be fleshed out by the member states.[8] Nevertheless, the requirement of minimizing the obstacles to internal trade and preventing the creation of new ones remains the frame of reference for the rulings of the European Court of Justice.

The validity of these and other[9] reservations held by commentators on the environmental dimensions of EU politics, should not make us lose sight of the historic process that is taking place: the development of political mechanisms that partly surpass the nation-state in an attempt to provide an answer to both the environmental side effects of economic globalization in an increasingly liberalized region and world and the internationalization of environmental problems in their physical and socio-economic dimensions. The principles and rules developed in the EU (or developed elsewhere (OECD, UNCED) but vigorously applied in EU legislation and policy making) alter the structure and functioning of market processes on a regional, that is supra-national, level in a number of distinct ways. The capital drain and other mechanisms by which the places with the lowest environmental standards and poorest environmental institutions used to achieve a preferential competitive position have largely been stopped since a kind of common environmental denominator has come about within the EU. Harmonization of environmental product and process standards within the European Union—or the future promise and perspective of harmonization—leads to homogenization of increasingly larger-scale production units across the continent and prevents the dumping of lower quality products or the concentration of (unpatented) old production processes in certain areas.[10] In the future political coordination and intervention may even result in more sustainable transport infrastructure and patterns, although at the moment decentralized economic and political decision making still dominates the developments in private and public transport.

Apart from these political and economic dynamics and developments, there is one other type of mechanism working against economic globalization from an environmental point of view: civil society and its vanguard, the new social movements. A European civil society can hardly be said to exist, as mentioned above. That means that the influence of all-European environmental organizations, such as the European Environmental Bureau (EEB), the European bureau of Friends of the Earth (FoEE), or the regional office of Greenpeace, have only marginal power to redirect the European integration processes into more environmentally sound directions. European media coverage, European public opinion (polls) or even the European Parliament, the national equivalents of which were so instrumental in the activities of national environmental NGOs in the 1970s, the 1980s, and the 1990s are relatively powerless. That means that only direct access to the European Commission remains as a viable European strategy for the Brussels based NGOs, besides of course the national routes. Although a "European" environmental movement is in the making, the national institutions and the European Council are still by far the most important strategic targets for NGOs in bringing about environmental change in Europe.

It has been argued that if a major European environmental initiative does not lead to parallel reforms and developments outside Europe it might disadvantage European capital in competition with other economic regions, most likely first in the triad. This would make such major European reform either ineffective or unenforceable over time. Although in most cases environmental costs do not form a significant part of the costs of products or production processes, as I will argue more extensively in the next chapter, they might make a significant difference in those cases where margins are small, where there is no future perspective of parallel reforms whatsoever (which will most likely be the case outside the triad), or where additional environmental costs can be substantial. This frames our analysis of the two other constituent parts of the triad, and the more global dynamics of trade and investments and environment within the MAI and WTO regimes.

Homogenization under Construction: The NAFTA Region

Before analyzing how in the countries that belong to the North American Free Trade Agreement increasing economic integration relates to environmental decay and environmental reform, it is essential to understand that NAFTA is fundamentally different from the EU. Although both are based originally on an economic treaty, the organizational structures, the policies of the two institutions, the supranational powers and the nature of the "common" markets differ fundamentally between the two. In analyzing developments within the NAFTA region, there would be little point in emphasizing the similarities and differences with developments in the European Union. If any comparisons are in order, NAFTA should be compared with GATT/WTO and MAI, although there is considerable disparity in scale, diversity of member states, organizational structure and the importance of environmental issues. However, the point in focusing on the NAFTA region is not so much to draw parallels and comparisons, but rather to understand how in this part of the global economic triad, economic integration and homogenization are accompanied by processes of environmental reform and environmental deterioration, and to identify the mechanisms of these "intermediate globalization" processes.

Among the three participants in the North American Free Trade Association (Canada, the United States of America, and the United Mexican States), Mexico is the economic and environmental outsider.[11] If it were not for the NAFTA, an economic analysis of this side of the triad would probably concentrate only on the former two. And if NAFTA was not there, one would expect diverging trends in economic development and environmental protection between the former two and Mexico. Yet, it is exactly the inclusion of Mexico in this regional analysis that makes it interesting for our purpose: to understand how globalization processes intertwine with efforts to safeguard environmental quality. And I will illustrate that especially the inclusion of Mexico in the trade agreement triggered the environmental debate on free trade and investment in this side of the economic triad and finally resulted in the greening of NAFTA.

The closer economic cooperation between the three states can be interpreted as a regional reaction to the European economic integration process and its main aim is without any doubt the reduction of trade and investment barriers and their further economic integration. It was believed that a regional free trade agreement would provide an additional impetus to economic growth, employment and prosperity and serve as an essential prop in the economic competition between the three regions of the global triad. Thus, if we look at the historic development of the North American Free Trade Agreement, it should not surprise us to find environmentalists among its fiercest critics, just like they opposed the GATT/WTO trade liberalization policies, the development of the MAI and economic integration and trade policies in the European region.[12] Government officials of the three countries (e.g., US EPA administrator William Reilly; US Trade Representative Carla Hills) as well as large numbers of NAFTA analysts consider the final outcome the "greenest" (multilateral) trade agreement ever concluded (Kumar et al. 1996; Vogel 1997; Hudson 1993; Johnson and Beaulieu 1996; Roberts 1998a; Hogenboom 1998; Esty 1999). But that does not automatically make NAFTA a pro-environment treaty.[13] NAFTA may come closest to what was concluded at the UNCED, but the trade and environmental objectives concluded in the 1992 Rio declaration and the related provisions of Agenda 21 fall short of what environmentalists would have preferred. As NAFTA is of more recent origin than the EU, I will focus on the process leading toward the emergence of NAFTA, the mechanisms that turned NAFTA "green," and a—primarily ex ante—assessment of the weak and strong environmental points of NAFTA.

The Emergence of NAFTA

The origins of NAFTA lie in the Free Trade Agreement (FTA) between Canada and the United States of America, which came into force on January 1, 1989. As there was relatively little trade between Mexico and Canada, Canada's attitude toward the NAFTA negotiations was rather defensive (Rugman 1994). The United States and Mexico were the two main propellers of the NAFTA negotiations that began in June 1990 and finally resulted in the signing of the treaty on 17 December 1992, entering into force on January 1, 1994. Distinct from the FTA, the NAFTA negotiations and the final treaty (including the associated North American Agreement on Environmental Cooperation, NAAEC) are characterized by a heavy emphasis on environmental considerations, which might justify its label "greenest" trade treaty. In a detailed analysis of the environmental debate that influenced the NAFTA treaty and its environmental side arrangement, Hogenboom (1998) has distinguished three phases, each characterized by distinct environmental policy topics, core actors, process dynamics, and final outcomes. During the first phase of the environmental debate, from the start of NAFTA negotiations in June 1990 until April 1991, the environmental effects of free trade and the scope for public participation occupied a central place in the environmental debate on NAFTA, with the unified environmental NGOs as the main participants. The second phase, May 1991 till September 1992, showed a remarkable sharpening in the discussions as the debate focused on two main items: the environmental provisions in NAFTA, and environmental cooperation between Mexico and the United States. While some environmental NGOs (NRDC, the Audubon Society, and the World Wildlife Fund) opted for environmental safeguards within the agreement, others (e.g., Friends of the Earth and the Sierra Club) pushed for an environmental charter, and NAFTA proponents found even that was going one step too far.[14] The 2000 *maquiladoras* in northern Mexico, the feared relocation of American industries to take advantage of laxer standards in Mexico, and the danger of downward harmonization of product standards were hot topics in this debate. Although not in principle related to NAFTA, the United States and Mexico negotiated an Integrated Environmental Plan for the Mexico-US Border Area. The rapid drafting, revision and conclusion of this plan show a concern on the part of Mexico about the consequences of its negative environmental reputation and publicity for the NAFTA negotiation process in the United States.[15] The final phase of the environmental debate on NAFTA, October 1992 to November 1993, focused entirely on the supplemental environmental agreement. As the final decision on NAFTA ratification was delayed until after the presidential elections due to a largely negative US Congress, it became an election issue. President Clinton's conditional support depended on environmental and labor side agreements, while Canada and especially Mexico were fiercely opposed to renegotiations and, especially, to opening up NAFTA again. While leaving NAFTA itself intact, the negotiations on the North American Agreement on Environmental Cooperation (NAAEC) centered on enforcement and sanctions with respect to sovereignty.

An Environmental Assessment of NAFTA

What was the final result of the NAFTA negotiations? If we are to assess the environmental gains of NAFTA we should look at both the treaty itself and the environmental side agreement. In the treaty itself (as well as in the route toward it[16]) several environmental innovations can be identified. (1) In the event of an inconsistency between specific trade obligations set out in specified multilateral environmental agreements and the NAFTA, the obligations in the former prevail. This explicit formulation runs against what is generally perceived as the dominant interpretation of GATT/WTO (Brack 1999).[17] (2) With respect to environmental standards, NAFTA clearly states that international standards are not "ceilings," but that member states can establish whatever levels of environmental protection they consider necessary, provided that such measures apply equally to domestic and foreign investors. (3) While GATT and the WTO protect countries from an unwanted influx of polluting industries, NAFTA's investment chapter to some extent prohibits countries lowering their environmental standards to attract polluting industries.[18] But perhaps the most innovative step can be found in the side agreement that is not officially part of NAFTA. The supplemental environmental agreement establishes a commission on environmental cooperation (CEC), headed by a council of senior environmental officials of all three countries, supported by a secretariat and advised by representatives of environmental organizations. Not only states can raise complaints, citizens also have the right to make submissions to the commission on any environmental issue. Moreover, the commission was given the authority to review the environmental implications of both production processes and products, diverging from WTO's limitation to products (chapter 7 below). In the enforcement procedures, designed to safeguard effective national enforcement of domestic legislation, the commission is also authorized to use trade sanctions and fines in cases of non-enforcement of new or existing environmental laws. These significant institutional improvements have made some scholars rather optimistic (Johnson and Beaulieu 1996; Audley 1997), and even led them to call the agreement a major move toward regional governance (Munton and Kirton 1994). Despite these environmental innovations, several limitations are regularly mentioned: the Secretariat does not have major investigative powers, it is strictly controlled by the Council and thus by the NAFTA parties, and the procedures for public complaints are different from those for complaints coming from NAFTA parties, to the detriment of the public (Kumar et al. 1996; Mumme and Duncan 1996). In addition, the CEC has performed successfully in conservation issues (birds, butterflies biodiversity) and some brown areas (sound management of chemicals, continental pathways of air pollution, transboundary impact assessment). However, its record on environmental standards, enforcement and pollution prevention has been rather poor and highly symbolic up until the late 1990s, while it has kept far removed from the more controversial "trade and environment" issues, and from issues regarding Mexican environmental performance.[19] The NAFTA Free Trade Commission and the CEC have not met in 6 years (Sanchez 2000). Moreover, NAFTA only provides possibilities to upgrade international standards into more environmentally sound standards, but no incentives to actually do so. Much to the disappointment of more critical environmental groups, sustainable trade and investment has not been put on the agenda via the NAFTA. Or as Johnson and Beaulieu (1996, p. 250) correctly put it, the NAFTA treaty is "rooted in the belief that the impact of environmental policies on North American trade are more worrying than the impact of trade policies on the North American environment."

If we are to analyze the dynamics behind the "greening" of this trade treaty, it is striking that it is especially the inclusion of Mexico—a country traditionally considered to be lax in environmental standards and enforcement but geographically close enough to both influence the American environment and profit from the American market—that provided major stimulants for a greening of the agreement in the early 1990s. The rather poor support of NAFTA in the United States and the existence of a considerable group of pro-free-trade and pro-environment congressmen and senators gave major opportunities to environmental (and to a lesser extent labor) interest groups to dominate the NAFTA debate and influence the negotiations. US environmental officials even participated in several of the economic working groups dealing with specific trade issues during the NAFTA negotiations. Although Mexico—the country that would feel the consequences of the environmental considerations in the NAFTA and NAAEC the hardest—was not very willing to hand in some of its sovereign powers on environmental policy and enforcement, the economic advantages were seen as too large to really obstruct negotiations and turn down environmental revisions.[20] At the same time US and Mexican environmental groups succeeded in collaborating during part of the process (although a major divide took place in the US camp; see above). According to David Vogel (1997) it is especially the fact that some environmental organizations were willing to support free trade in exchange for specific provisions in the side agreement which makes the NAAEC more powerful than its labor counterpart (NAALC). It was in fact the constant quest for environmental homogenization that united environmental NGOs and some traditional political actors (EPA, congressmen), while they diverged in their strategies to use NAFTA to ecologize North American economic production, consumption and trade. The crucial role of active (especially US) environmental NGOs in greening the treaty contradicts the loss of interest of these environmental NGOs during the implementation of the NAFTA treaty, but explains to some extent the limited actual environmental record of NAFTA (Sanchez 2000). The weakening of the CEC can be partly explained by the retreat of the environmental NGOs.

The question is what difference these environmental provisions make in practice. Several authors have done some initial analyses of the work of the various organizational structures under NAFTA and NAAEC, showing that the environmental records are indeed not overwhelming. The trade community managed to strengthen its role during NAFTA implementation, allowing the governments to interpret the environmental provisions in a minimal way. The lawsuit by the Ethyl Corporation, American-owned producer of the gasoline additive MMT, against the Canadian government gives evidence of potentially dangerous environmental consequences of NAFTA. After unsuccessfully trying to dissuade the Canadian parliament from banning MMT with the threat of a lawsuit, the Ethyl Corporation finally managed to convince the court that Canada's legal ban on import and transport of MMT within Canada constitutes an illegitimate expropriation of the firm's assets. Ethyl Corporation received US $13 million compensation from the Canadian government (although it sued for US $251 million originally).[21] But it is too early to make a definite judgment. According to Vogel (1997, p. 359) "it is worth noting that it took the European Community nearly a generation to begin a serious effort to improve the environmental standards of its southern, less "green" member states. It may well, therefore, be several years before NAFTA has a significant impact on Mexican environmental standards." NAFTA was able to prevent a "race to the bottom" as Mexico's increased economic dependence resulted in a strengthening of its environmental policies rather than a weakening. In addition, the influence of the environmental provisions in NAFTA and its side agreement will most probably reach beyond the three countries and will influence future international or global trade policy making in the WTO.[22]

Changing Modes of Governance?

The CEC has not yet turned out to function as a regional form of environmental governance, as a predecessor of an EU-like institution with supranational powers. Rather, much of the reluctance of the parties to give the CEC major powers is rooted in the parties' defense of their sovereignty and their unwillingness to create an image of a supranational CEC. If sovereignty was undermined during the NAFTA process it was rather via the empowerment of non-state actors. During the creation of the NAFTA treaty it was especially the environmental NGOs that managed to secure a major role in environmental politics. The lack of interest, commitment and involvement of these NGOs in the implementation phase raises questions about the role of these organizations in environmental governance. In the implementation phase, the market actors managed to challenge environmental regulations without political supports from their national governments, especially via the dispute settlement scheme. Morales (1999) has called this the relocation of authority competencies from state-centered actors to non-state actors and the creation of a hybrid model of governance.

A Region of Nation-States: Japan and Newly Industrializing Asia

The third side of the global economic triad is located in East and Southeast Asia, with Japan as the economic superpower and a growing number of first-generation (Hong Kong, Singapore, South Korea, Taiwan) and second-generation (Indonesia, Malaysia, Thailand, Vietnam, China) newly industrializing economies closely related in terms of international trade and investments. Most economists analyze growing economic integration and growing linkages of trade and investment between the East and Southeast Asian economies. Japan, Korea, Taiwan, Singapore, Malaysia and Thailand, in particular, show major levels of FDI outflow toward other East and Southeast Asian countries from the late 1980s on. But if I compare this side of the triangle with the other two at least two important differences should be noted:

• The economic interdependencies in trade and investment—though growing—are still relatively slight in Asia (Burnett 1992; ADB 1996; Castells 1996) and contain greater inequalities, as compared to the other two economic blocks of the global triad. Or, as Lloyd (1994, p. 135) concludes in his study on regional integration: "East Asia is much less integrated than the European Communities or North America, in terms of the degree of uniformity of policies and of prices." In contrast to the EU and NAFTA, growing economic integration takes place less via the building of external walls and barriers in favor of the internal regional developments, but rather via "open regionalism" and market-led integration (Higgott 1999).

• Homogeneity between Southeast and East Asian countries seems to be much lower than in the countries of the other two parts of the global triad. Although, as we have seen, NAFTA also contains quite distinct member states in terms of economic development, environmental protection levels and policy culture (to name but a few relevant characteristics), the political and economic dominance of Canada and of course especially the United States over Mexico is quite different from Japan's hegemonic position with respect to the newly industrializing countries and peripheries in that region, not to mention these countries' historical, geopolitical and demographic relationships with Japan and China.

These differences contribute to distinct processes of regional environmental reform in connection with globalization in the Asian region, as compared to the processes evolving in Europe and North America. But the idea that this region can hardly be seen as homogeneous should be balanced.

Notwithstanding the differences in, for instance, levels of economic development, FDI, and industrial structure between the various (first- and second-generation) "tiger" economies and Japan, something special seems to be the case in the Southeast and East Asian region. In his *opus magnus* Manual Castells (1997b, pp. 206–309) signals the coming of a Pacific Era: the era in which the pacific region will appear to be the main center and trigger of economic innovation, transformation, and development in the world. Although he concludes that there is no institutional, cultural, political, or even economic Pacific region (in the sense of shared or common characteristics),[23] the "similarities" of the development process of the countries in this region lie in the decline of Western economic and technological dominance and in some parallels in state-economy relations between the Southeast and East Asian developing economies. The most significant of the commonalities, in Castells's view, is the role of the state in the developmental process. As argued extensively by Evans (1995), these states had, and still have, the political capacity and relative autonomy to impose their project of a strong state on society, coupled with high levels of economic growth, low wages, limited internal democracy and the construction of a national identity. At the same time these states did not have to make too many concessions to society. The so-called external networks of the state, connecting the state with civil society and market parties, are essential here. The autonomy of the state in Southeast and East Asia is embedded in external networks with crucial economic and societal actors, although the nature of the embeddedness—the characteristics of the ties between the state and the surrounding social structure—differ somewhat between individual countries.[24] The effectiveness of the state emerges not only from its internal bureaucratic capacity and coherence but also from the complexity and stability of its interactions with market players and civil society organizations. A strong "developmental state" seems a crucial factor in the economic success of both Japan and the Southeast and East Asian "tiger" economies.[25]

Japan: Environmental Transformations in a Developmental State

There is some empirical evidence that such developmental states with strong links with society are not only able to pursue a successful industrialization paths or transform traditional industrialization patterns to fulfill the requirements of reflexive modernity (or what Castells calls the Information Age). They also have a good basis for effectively and rather quickly reforming industrialization patterns into more environmentally sound directions. Japan, Singapore and, to a lesser extent and more recently, Taiwan (Rock

1996a) have—after a period of severe deterioration—radically reduced the environmental impacts of industrialization in a short period of time. In his extensive study on environmental politics and protest in Japan, Broadbent (1998) has analyzed the dynamics behind the relatively early and far-reaching environmental reforms in Japan, and the stagnation in environmental innovations from the late 1970s until at least the late 1980s. The closely interrelated ruling triad of the state, the dominant LDP party and the major business groups, with major influence in the various sections of society, was able to accelerate institutional reforms for the environment in an early stage after fierce societal protests following the Minemata and Itai-Itai epidemics, among other things. The structural changes in Japan's economy and its related environmental performance, as reported in studies by Jänicke et al. (1990; 1992; 1995) originate in this period. But the ruling triad in Japan was equally capable of blocking further environmental innovations later on (Maull 1992; OECD 1995; Broadbent 1998). Major national environmental NGOs aiming at structural economic reform never found fertile ground in Japan for longer periods, due to (according to Maull and Broadbent) both these institutional characteristics of the "developmental state" with its strong societal ties, and the prevailing culture of non-resistance against the government. Although environmental protests did manage to move the ruling triad to undertake major environmental reform programs and institutionalize environmental protection, the initiative was soon taken over by the "extended state," while the more business-oriented factions (Ministry of International Trade and Industry, MITI; business groups; and parts of the LDP) managed to slow down the radical parts of the reforms after the mid 1970s. By then environmental struggles increasingly concentrated within the ruling triad, where the Environmental Agency and the environment-oriented parts of LDP lost some of their power and environmental innovations stagnated, although most of the major steps forward were not reversed. It was not until the late 1980s and the early 1990s when the emergence of cross-border and global environmental issues, the ensuing international environmental criticism of Japan, and the increasing globalization of the Japanese economy resulted in an upsurge of environmental consciousness and a Japanese "revival in environmental policy developments" (OECD 1995, p. 31; see also Maull 1992), both in the national and international arenas.[26] Japan's environmental record in Multilateral Environmental Agreements, domestic politics and ODA has since improved considerably. Although public support for environmental NGOs increased in the early 1990s, they were unable to use this support strategically to accomplish far-reaching environmental reforms (Mitsuda 1997). According to Maull (1992) it was especially the transnational environmental NGOs and a number of developed nation-states that put effective pressure on the Japanese government to be more serious about domestic and international environmental policies.

Japan's Environmental Reform in Regional Perspective

A major part of the successes in environmental performance of the Japanese economy is caused by the structural changes in its economy, with a shift from heavy, and heavily polluting, industries to processing and assembly industries and services. Some of these heavily polluting industries re-emerged in the newly industrializing countries (NICs) in Southeast and East Asia, financed either by local investments or by Japanese FDI (McDowell 1990; Burnett 1992). Analyzing two waves of Japanese foreign direct investment in Southeast and East Asia, Cameron (1996) is quite skeptical and pessimistic regarding its contribution to ecological improvements in industrial developments in these NICs. The first wave of Japanese FDI in the 1970s can be explained by internal developments in Japan; it was basically the industrial practices that were no longer tolerated in Japan—or could not compete—that were relocated to other Southeast and East Asian countries, taking advantage of weak environmental and other social regulations and giving little consideration to the local environment. The second wave in the late 1980s is similarly analyzed by Cameron in terms of what might be called a "colonization of Southeast Asia's environment" or Japan's "ecological shadow" (Maull 1992), although the direction of FDI changed somewhat to the tourist sector and continued in the exploitation of natural resources such as tropical forests and mining. Despite the impact of foreign industrial investments on the local environment as analyzed by Cameron, Wallace (1996, p. 68) argues that these investments also introduce modern production technologies into these NICs that generally perform better in environmental terms than the domestic industries of these newly industrializing economies themselves. This better environmental performance per unit of economic output is explained by the growing importance of the economic fact that pollution generally equals a waste of resources (the "pollution prevention pays" principle), by the fact that FDI is put primarily into the manufacture of products for a global market that requires standard (also environmental) product quality and, hence, refined production technologies, and by the fear among international industries of public censure (or even legal action) at home for their environmental misbehavior in developing countries.

Some of these characteristics of Japanese environmental reforms can also be spotted in other Asian states (O'Connor 1994; Rock 1996a; Angel et al. 1998), though colored by their national characteristics. These countries were neither in the position to simply "export" their most devastating production processes to their regional peripheries or extract natural resources from these peripheries as easily as Japan was able to do for considerable time (Burnett 1992).[27] Consequently, the delinking of economic growth from environmental parameters that was evident in Japan and the EU, is almost absent in these "tiger" economies, as pollution intensities of GDP increase (Angel et al. 1998; but see Sonnenfeld 2000 for some contrasting evidence). In none of the countries in this region have environmental NGOs reached the level of the United States and Europe in terms of members and influence and in most countries their room for maneuver is significantly constrained by the state (Mittelman 1999). But there are differences within the region (e.g., Thailand's active environmental NGOs), and the trend is one of an emerging role of civil society agents pushing for environmental reforms against the intricate connections between the state and economic powers (Eccleston and Potter 1996; Sonnenfeld 1996; Clarke 1998; Mittelman 1999; O'Rourke 2000). Similarities and differences are also visible in the specific networks of (economic and political) power that facilitate economic development and industrialization. These and other differences notwithstanding, the common characteristics of developmental states lend this region a certain degree of homogeneity. Global developments beyond this region also work toward environmental homogenization: global convergence of environmental standards; international cooperation between national environmental agencies; global cooperation of environmental NGOs; global green markets; TNCs setting and implementing global environmental standards and practices; and green supply chains reaching around the world. All these factors have a positive and to some extent harmonizing effect on environmental practices in the Asia Pacific region. But this homogenizing trend has not yet resulted in strong regional integration, either in terms of economy or in terms of the environment, nor in significant developments toward powerful regional institutionalized governance.

Regional Organizations for Economic Integration: ASEAN and APEC
In contrast to the other two sides of the global economic triad, the Asian side does not really have any strong regional treaty, organization or institutionalized form of governance that covers both global economic development and environmental protection or reform. Among a large number of regional (economic) forums and organizations, there are two more comprehensive and potentially important regional economic organizations in that part of Asia: the Association of Southeast Asian Nations (ASEAN) and the Asia-Pacific Economic Cooperation forum (APEC).[28] Although these organizations are both primarily focused on economic integration and to a lesser extent political cooperation, they do contribute to the development of a sense of regional identity, as Higgott (1999) rightly argues. This sense of regional identity in the making is slowly becoming a factor in interregional relations, giving integration attempts a new non-economic dimension. Although there exists no collective objective comparable to a European Ideal, the Southeast and East Asian region is more than a set of nation-states linked only by trade and investment. ASEAN and APEC can also be seen as "materializations" of this growing regional identity.[29]

ASEAN is an organization of developing and newly industrializing countries in the region, established in 1967 to promote political and economic stability and to create—especially after 1975—a united front against the emerging communist states (Plummer 1997). Economic integration in ASEAN began to deepen in the late 1980s, culminating in the ASEAN Free Trade Agreement and ancillary agreements in the early to mid 1990s. AFTA was concluded in 1992, mainly as an instrument to attract larger FDI to the region, at a time when competition for FDI was increasing (for instance with China). In contrast to NAFTA, for instance, AFTA should be interpreted as an outward-looking economic integration process, closely linked to domestic policy reforms in the "member states." Most of the barriers to import have been reduced unilaterally. As Ariff (1996, p. 218) states "ASEAN has carefully avoided integration schemes that would reorient member economies. . . . A laudable feature of AFTA is that it is aimed not at increasing intra-ASEAN trade but at making ASEAN products competitive in the world market and making the ASEAN region attractive as a center for FDI." ASEAN and AFTA have devoted little attention to environmental issues as related to trade liberalization and international investments programs, quite in contrast to the institutions in the other two sides of the triad. A plan of action on the environment—aimed at strengthening legal and institutional capacities and harmonizing air and water quality standards and EIA procedures—and a working group on transboundary pollution are the main environmental highlights. ASEAN has largely kept aloof from civil society activities, denying the few regional environmental organizations such as the Climate Action Network access to ASEAN meetings and negotiations (Mittelman 1999, p. 84). And lastly, the hegemonic economic power of the region, Japan, is not a member of ASEAN or AFTA, which on the one hand makes them more homogeneous, but at the same time less strong if they are to function as the economic integration institutions of this part of the global triad.

APEC is another matter, however. Established in 1989, APEC shares with NAFTA the fact that among its 21 member states there are both developing and developed economies in the Asia-Pacific region, including the United States, Canada, Australia, and Japan.[30] Its core fields of cooperation are trade and investment liberalization and facilitation (open regionalization) and development cooperation, and it operates on the basis of consensus rather than binding agreements such as NAFTA or the EU treaty. The degree of economic integration is rather marginal up until now and environmental issues are even more peripheral in APEC than in ASEAN/AFTA. Beginning in 1993, APEC has given some attention to marine resource conservation and fisheries and more recently it has identified cleaner industrial production as a relevant issue, but these aspects are only marginally integrated with economic integration policies on trade and investment. The APEC sustainability agenda remains "a flurry of small-scale, capacity building projects and little else beyond statements of principles" (Gershman 1998) and several environmental NGOs from various countries have recently criticized APEC for being "anti-democratic, unaccountable and untransparent" (as quoted in Mittelman 1999, p. 85). Although Anderson and Brooks (1996) see APEC as an obvious forum for discussing relationships between environmental measures and liberalized trade and investment, and integrating them in that region, hardly any environmental initiative has been developed to date. Japan and the United States would be the most obvious candidates for triggering these regional environmental initiatives, both being environmental front-runners from the point of view of their domestic policies. Maddock (1995), for instance, believes that only Japan is in a position to articulate the environmental security interests of the region and create a regional system of environmental management. It should not surprise us that, despite the poor environmental performance of APEC, as long as there is no institutional basis for an integration between economic globalization and environmental reform, domestic environmental NGOs will hardly look beyond the border of their nation-state (let alone that significant and powerful regional NGOs will emerge). Compared to the EU and NAFTA regions, the role of Asian environmental NGOs in greening regional economic globalization beyond the nation-state is marginal.

Environmental Reform in Asia

In conclusion, environmental reform in the East and Southeast Asian region does not seem to work via any supranational or intergovernmental organization. The existing regional institutions and organizations are still too weak and hardly committed to the environment, although the major 1997 Indonesian forest fires that affected Singapore and Malaysia seems to have triggered some regional environmental cooperation. The main factors in regional environmental reforms are not specifically regional but either national or global: global environmental regimes (e.g., the International Tropical Timber Organization, the Montreal Protocol under the Vienna Convention, the International convention on toxic trade, the Biodiversity Convention); strong national environmental NGOs in some countries—Japan, Taiwan, Thailand, Indonesia—with links to global activist networks; support given by international development assistance programs to newly industrializing economies (e.g., by Japan's JICA, but also programs run by the EU, the United States, Canada and Australia), often accompanied by environmental goods and services exported by the donor country; global green markets that push regional or national producers into environmentally sound production; and, occasionally, environmentally forward-looking TNCs.

The Triad . . . and the Rest?

This chapter began with the claim that economic globalization is largely restricted to the global triad of the European Union, the NAFTA region, and Japan and the emerging "tiger" economies of Asia. Although this claim still stands, it should not be taken to imply that globalization, and the concomitant environmental deterioration and reform, do not affect the societies outside the triad. As various scholars (e.g., Sachs et al. (1998)) have convincingly argued, some of the most severe side effects of economic globalization are felt outside the core economic regions. Environmental devastation, declining sovereignty, unequal distribution and other consequences of global capital movements are also, or perhaps primarily, found in the peripheral regions, especially since—in varying degrees—these societies are often deprived of the benefits of economic globalization and liberalization. In a similar way, environmental reform connected to globalization can only be understood if the specific position of "peripheral" countries is taken into consideration. Although there is a shared concern among all nation-states that progress in environmental protection is both necessary and desirable, there is apprehension, especially among the developing countries, that rules formulated in agreements on environmental protection by WTO, NAFTA, the EU, and various multilateral environmental regimes could be used as a disguised form of protectionism. Such apprehension is even felt toward unsuspect technical regimes, such as that of the International Organization for Standardization.

7
Global Environmental Reform beyond the Triad

As I have indicated in general terms in chapters 3–5, processes of globalization do not leave the national regimes of environmental reform unchallenged, either in the developed nations (as indicated in chapter 6) or in developing countries. But the national environmental regimes of triad countries are connected to globalization processes in a different way than those of peripheral countries. First, though regional and global environmental treaties, institutions, norms and principles do have a homogenizing effect on various countries' environmental regimes, not all countries have equal opportunities, possibilities, and interests to fulfill these homogeneous demands. Second, it has been proved time and again in practice that the different sovereign states that make up the international system of nation-states do not have equal access to the design and functioning of international regimes, regardless of the principles of equality. That means that the triad countries, with better resources, generally play a larger role in designing global regimes than peripheral countries, and as a result international institutions serve their interests better. Third, owing to various economic, political, and "cultural" mechanisms, the construction and the transformation of national environmental regimes in one part of the world are bound to affect such regimes in other parts of the world. Here, too, the countries of the triad are usually in a better position to influence the regimes of developing countries than vice versa.

In this chapter I focus on global environmental reform in between the triad of the EU, the NAFTA region, and Japan and newly industrializing Asia, on the one hand, and the developing countries in the South, on the other. I will concentrate on two key global institutions (the Multilateral Agreement on Investment and the GATT/World Trade Organization), on global environmental regime formation via multilateral environmental agreements, and on pollution relocation and the controversy over the environmental Kuznets curve.

Investments, Trade, and Environment: MAI and WTO

Besides the environmental consequences of regional economic integration in the three centers of economic globalization, we also have to analyze the environmental dimensions of interactions that surpass each of these regions and in fact strongly connect the economic triad with other parts of the world. The World Trade Organization is without any doubt the most important global political-economic institution, believed by many scholars to be a constant threat to the environment and a hindrance to global environmental reform programs. However, before concentrating on the WTO, I will examine the OECD initiative to reach a Multilateral Agreement on Investment. This MAI is to a major extent complementary to the WTO, as the latter concentrates on world trade (although it is moving into the direction of the relationship between trade and investment), while the MAI is basically focused on regulating—or deregulating—global investment in all economic sectors.

The Multilateral Agreement on Investment

Bilateral investment treaties between states have a long history and have grown spectacularly in recent times. This profusion of bilateral agreements was increasingly felt to be confusing and potentially inefficient. Several attempts have been made to arrive at a multilateral approach (Schrijver 1995), and the OECD has undertaken the latest and most comprehensive attempt.

After an OECD ministerial council meeting in May 1995, a working group was set up to develop a proposal for a Multilateral Agreement on Investment, aimed at both protecting transnational investments and liberalizing global investments. The increasing amount of foreign direct investment by OECD member states can be seen as a stimulus to the development of such an agreement. Initially, the developments and negotiations regarding the MAI went on rather secretly, with hardly any discussion or information disclosed to the public and or to the non-member states. Only toward the end of the negotiating and drafting process among the 29 OECD member states did the MAI begin to attract attention among the public at large, and only then did the debate over its various consequences gain momentum.

Beginning at the end of 1997, the MAI came under increasing criticism from environmental NGOs, from labor, consumer, and Third World solidarity groups, and from developing nations.[1] This growing "political backlash" against the MAI in one country after another resulted in a decision by the OECD Ministerial Conference in April 1998 to temporarily suspend negotiations for "a period of assessment and further consultation between the negotiating parties and with interested parties of their societies."[2] The OECD ministers called for "a transparent negotiating process and an active public debate on the issues at stake in the negotiations." After 6 months, just when the negotiations were about to be resumed, France concluded that the negotiations had failed and refused to send a delegation to the final OECD meeting. As a result, the October 1998 meeting of the OECD could no longer conclude the MAI negotiations with an agreement. The threats to its national cultural industry were the prime motive for France to make this move, but environmental considerations—and especially those put forward by the NGO coalition against the MAI—also played a significant role. Most probably, however, this delay will not mean the end of the MAI. Some expect negotiations on the MAI will continue within the OECD framework. France, the European Union, and other countries would prefer (and are actually trying) to renegotiate within the WTO framework, giving developing countries a better position.

The Multilateral Agreement on Investment is the first attempt to combine three areas of regulation with regard to foreign direct investment: investment protection, investment liberalization, and dispute settlement. Backed strongly by the United States and the European Union, the MAI is based on the NAFTA investment provisions but amplifies them and extents their geographical scope. The objective of the MAI is to create a "level playing field" with uniform rules on market access and legal security and without distortions in and restrictions on investment flows. In that sense, the emergence of the MAI is a typical product of the globalization era, regulating on a global scale what nation-states used to regulate on the national level. In principle, the goals of the MAI are hardly in dispute: regulation of the growing transnationalization of investments and equal treatment in each country of domestic and foreign investments. The agreement entitles foreign companies to buy, sell, and (re)locate plants, resources, assets, and other properties on the same legal conditions as domestic firms. But the procedures and instruments included in the MAI to accomplish this goal are criticized for various reasons. Developing and transitional countries, for instance, have criticized the MAI for having been designed for the benefit of the well-established and developed nation-states. In their view, the MAI obstructs diversification of their former economies and requires a level of liberalization that does not fit economies in "construction." In addition, withdrawal conditions rule out temporary tryouts. Although these countries will face strong economic pressure to sign the MAI, all objections center on the restrictions placed on the powers of national governments to restrict foreign investments and pursue an economic policy in what they consider to be national interest. Virtually no "exceptions" or "reservations"—other than for military or security reasons—are allowed for protecting the national culture, public health, social rights, or the environment.

Environmental criticism of the MAI—which is our prime concern here—focuses on the restrictions placed on the environmental sovereignty of nation-states by the following two absolute principles: that foreign private ownership cannot be made subject to any general limitation or set of conditions and that compensation is to be paid for what are called "expropriations" (whether due to nationalization or "any other measure with such an effect"). Although it is prohibited for governments to discriminate against foreign investors, there would be nothing to stop governments from offering foreign corporations more favorable conditions than domestic ones, for example by setting less stringent environmental requirements. The MAI is thus often seen to contribute to the "race to the bottom," since the need to attract globally flowing capital will put pressure on states to lower environmental standards and soften regulatory regimes. According to the dispute-settlement provisions, private investors can bring national governments before an international court. This is a novelty in international agreements (NAFTA allows such investor-to-state dispute resolution in limited cases, but not before international courts) and an encroachment on the sovereignty of nation-states. According to an international coalition of environmental, consumer, and development NGOs, this is an agreement that regulates governments rather than international investment. Conversely, private investors cannot be made to behave according to international standards on environment and labor conditions. And consequently, states cannot bring private firms to the international court, since private companies have no obligations, but only rights. In short, opponents claim that the rights of investors are well looked after, while the MAI imposes virtually no responsibilities on investors regarding environmental protection and other issues.[3] In the global battle between the international system of sovereign states and the transnational companies, therefore, the MAI definitely favors the latter.

The potentially disastrous effects of the MAI have frequently been pointed out with reference to legal procedures that have occurred in the framework of two other economic agreements: the MMT case of the Ethyl Corporation versus the government of Canada under the NAFTA regime (chapter 6 above) and the WTO's recent decision against the European Union's ban on hormone-fed beef imports from the United States (January 16, 1998).[4] Neither of these was a case of discrimination against foreign producers as such, but still they were instances where the national or common environmental and health safeguards were lifted for the benefit of economic producers.

The MAI was turned down primarily by France, in an attempt to protect its cultural industry and cultural heritage. Meanwhile, the global environmental movement—and other parts of the opposition to the MAI—managed within a short period of time to generate a great deal of high-quality information on the World Wide Web, mobilizing important and influential sections of the global civil society and thus succeeded in putting several governments under severe pressure (Esty 1999b). The Canadian Environmental Law Association[5] argues that two reports were especially influential in stopping the MAI. In her report to the French government, Catherine Lalumiere, a member of the European Parliament, stated that the NGO opposition to the MAI went to the heart of the agreement, since it posed a serious threat to the sovereignty of nation-states. The report demanded seven fundamental amendments to the MAI and is believed to have been influential in the withdrawal of France. The second report by Jan Huner, Secretary to the OECD Chairman of the MAI negotiating group, commented on the meeting between MAI negotiators and environmental representatives of October 1997. Huner concluded that the meeting was "decisive because some of the points raised by environmental groups convinced many [negotiators] that a few draft provisions, particularly those on expropriation and on performance requirements, could be interpreted in unexpected ways" and that "the dispute between Ethyl Corporation of the US and the Canadian Government illustrates that the MAI negotiators should think twice before copying the expropriation provisions of the NAFTA."

The stream of increasingly firmer demands for major changes in the original draft contributed to a growing uneasiness among the MAI supporters and will prove to have influenced the MAI draft considerably once negotiations recommence, within either an OECD or a WTO framework. Still, like the WTO, the MAI will remain primarily a global economic agreement in which environmental side effects can only be minimized at best. It is not likely to develop into an instrument for multilateral environmental reform.

GATT and the WTO

The General Agreement on Tariffs and Trade (GATT) differs on various accounts from the MAI. Although both aim primarily at economic liberalization and at eliminating barriers and distortions that hinder the ongoing progress of liberalization, the history of the agreement, its institutionalization and organizational foundation, the members involved in negotiating and further developing the agreement, and the actual impact it has had on the global economy and global environment are quite distinct.

GATT came into force on January 1, 1948, originally as a kind of provisional agreement on some basic principles of international trade until the International Trade Organization would come into force. But mainly because the United States—the driving force behind this third Bretton Woods institution—began to withdraw its support in the late 1940s, the ITO was never established. Although GATT was never meant to stand without the ITO, it still survived. Only in 1993 was the World Trade Organization established, at the end of the Uruguay round of GATT negotiations. GATT's most fundamental principle is non-discrimination, consisting of two components: most-favored nation status (the obligation for each signatory to treat imported products from any other contracting party as favorably as "like" products from any other GATT member country) and national treatment (imported products should be treated no less favorably than "like" domestically produced goods once they enter the country of importation). The WTO serves as an umbrella organization supporting GATT and other elements of the international trading system. With the WTO coming into force, major changes have appeared in the dispute-settlement procedures of GATT (consensus is no longer required), although all major policy matters of the WTO still require consensus among its member states.

Although GATT does not mention the environment specifically, the heart of the GATT environmental debate is article XX, which lists the exceptions to the general rules and principles, including provisions for some environmental regulations. Article XX has been interpreted narrowly, limiting the scope for trade restrictions on environmental grounds. Environmental policies are justified only if more GATT-consistent alternatives to reach the same goal are unavailable. Conservation measures are not allowed if they aim to protect the global commons outside the jurisdiction of the nation imposing the measures, such as the oceans, the atmosphere and the ozone layer. In addition, trade restrictions on environmental grounds are permitted only if they refer to the physical or chemical characteristics of a product, not if they merely refer to how it is produced.[6] These last two stipulations seem to be remnants of a time when environmental problems were primarily defined in terms of local pollution issues and fully sovereign states, ignoring any transboundary or global dimensions and interdependencies (Vogel 1997).

Esty (1994) mentions three general objections raised from an environmental point of view to GATT and the present international trading system:

- GATT rules are unbalanced, giving more weight to market access than to environmental protection and leading trade principles to override legitimate environmental interests. If high environmental standards must be distinguished from "protectionism in green disguise," the outcome is too often to the detriment of the environment. In addition, a balance should be struck between the commercial advantage of uniform standards and easy market access on the one hand, and the environmental and economic benefits of regulatory diversity (i.e., standards tailored to local conditions and preferences) on the other. But within GATT the former prevails over the latter.
- Attempts to use trade sanctions and restrictions as leverage to enforce international environmental agreements and to discourage transboundary or global spillover are hardly ever condoned by the GATT regime. GATT—like the MAI—is only or mainly concerned with the problem of "high" environmental standards placing too great a burden on trade, and hardly concerned with "low" environmental standards that place countries under the burden of pollution externalities. Free riders on global environmental problems can easily obtain unfair trade advantages. The trade provisions of the Montreal Protocol do acknowledge this and ensure that no country can try to obtain competitive advantage on CFCs.
- Countries with lax national—now not international—environmental standards can obtain a competitive advantage in the global marketplace at the expense of the environment. GATT offers no possibilities for any unilateral drive to exact convergence or harmonization of diverging environmental regimes.

The question emerges how the primarily economic GATT/WTO regime relates to the transnational environmental regimes and policies that are founded in various multilateral environmental agreements (MEAs). Trade provisions in MEAs have been designed either to exercise control over trade itself (e.g., CITES, the Basle convention) or as an enforcement measure to make sure that non-parties do not undercut the MEA (e.g., the Montreal protocol). Up to 2000 there had never been a dispute case involving an MEA trade measure before a GATT or WTO panel or Appellate Body, and according to Brack (1999) it is uncertain how such a panel would rule. Neither—as international lawyers consistently argue—is there any legal principle or international law that defines a hierarchy between the two sets of international agreements.[7] But in the daily practice of international environmental politics it is often found that GATT/WTO rules take precedence above those laid down in multilateral environmental agreements. The perception that certain parts of MEAs (in drafts or amendments) may be incompatible with GATT and the WTO results in a "political chill": some parties refuse to adopt or to ratify such MEA provisions.[8] It is basically the structural characteristics of the international system and organizations that results in such divisions of power.

"Greening" the Trade and Investment Regimes

The meager possibilities of dealing with environment and trade issues within GATT and the WTO and the observation that multilateral environmental agreements are often weaker than the powerful trade and investment regimes are two important insights for the discussion on how to strengthen global environmental politics that link to global economic liberalization. The three basic strategic alternatives advocated in this debate are to build environmental considerations into the existing trade regime, to develop an independent and powerful environmental organization and regime that can counterbalance the WTO and the MAI in global environment-and-trade/investment issues, and to continue with the construction of a global environmental regime along the present line of piecemeal multilateral environmental agreements that together turn (economic) globalization patterns into more environmentally sound directions.

One of the major complaints voiced at the finalizing of the Uruguay round of GATT negotiations concerned the limited attention paid to environment-and-trade issues. Consequently in Marrakech, at the concluding meeting following this Uruguay round of negotiations, it was decided to establish a trade-and-environment work program and a

Committee on Trade and Environment to begin working on this issue in preparation for a new round of negotiations. This very same committee concluded, however, that the WTO is not very likely to become the main international designer and enforcer of environmental rules, since "the competence of the multilateral trading system is limited to trade policies and those trade-related aspects of environmental policies which may result in significant trade effects for its members" (WTO-CTE 1996).[9] This is not to say that GATT and the WTO do not require any "greening," as the present environmental provisions cannot even fulfill a limited role in leveling environment and global trade. Several minimal and more far-reaching improvements on GATT procedures and contents have been suggested from different sides, including making procedural changes in the dispute-settlement procedures, hiring additional environmental staffing at the WTO, conducting environmental assessments of any substantial trade negotiation, open processes regarding decision making and information dissemination, designing a new WTO Agreement on MEAs similar to the ones on Agriculture and on Technical Barriers to Trade, setting standards for assessing the legitimacy of environmental trade measures, and introducing a trade balancing test in article XX (Brack 1999; Wallach and Sforza 1999).

There are indeed some indications that the WTO is becoming more sensitive to criticism from an environmental point of view and other protests and arguments. During the preparations for the Seattle meeting in November 1999, high-level WTO officials made it clear that they saw the organization's negative environmental image as a major concern. Also, the fact that multinationals were given preferential access to this Seattle meeting (at the initiative of the main organizer of the meeting, Bill Gates) was sharply criticized by both the EU Commission and the president of the United States. By contrast, the Netherlands' official delegation to Seattle included a representative of Greenpeace as well as a representative from the country's main business association (VNO/NCW). Apparently, there is a growing conviction that issues of trade liberalization can no longer remain restricted to a select audience of "technical" negotiators, politicians, and business lobbyists.

But substantial "greening" will prove difficult, as Esty (1994, p. 206) summarizes, owing to fundamental international differences of opinion on "the appropriate trade-off (if any) between economic growth and environmental protection,[10] the changing nature of sovereignty, the use of power in the international realm, the optimum structure for democratic decision-making, and the proper role of interest groups in governmental processes." And indeed the November 1999 WTO summit in Seattle showed a variety of problems such as these. Meant to set the agenda for a new round of trade liberalization, the Seattle summit failed owing to a number of problems, including differences of opinion on agricultural liberalization (in relation to agro-environmental issues such as multi-functional agriculture and beef containing hormones), major and well-organized criticism coming from both reformist and radical environmental and other NGOs (both at a distance and in Seattle), poor preparation, and an American chairman who seemed excessively influenced or even dominated by domestic business interests. Any "greening" will always remain a compromise within what is predominantly an economic trade regime.[11] The maturation of the WTO into something similar to the European Union, acquiring sufficient supranational powers, democratic properties, and environmental as well as economic objectives, would be a more promising strategic trajectory, but even more difficult to accomplish.

Not surprisingly, the idea re-emerges at regular intervals that, instead of attempting to integrate the two objectives into one global organizational framework and regime, it is more rewarding to set up a separate international environmental organization with enough power, mandate, and competence to counterbalance GATT and the WTO in economic globalization processes. Few see the possibility of building such an institution from the existing United Nations organizations, UNEP being the most likely candidate. A new global environmental organization, it is felt, should formulate the basic principles of global environmental policy making from an "ecological rationality" perspective parallel to what GATT and the WTO have done from an economic perspective. Only the construction of such a parallel institution can lead to a more balanced integration of economic and trade liberalization (on one hand) and environmental protection (on the other). In such a perspective, the common principles underlying an increasing number of multilateral environmental agreements can be combined into one common organizational and legal basis, thus improving the consistency of this piecemeal regime development. It goes without saying that for many scholars such a strategic goal is "beyond imagination,"[12] and that for some it is undesirable. But if we were to take the trends of ongoing global economic integration seriously, an ecological modernization perspective would guide us in this direction.

Global Environmental Regimes and Non-Triad Regions

The struggle to "green" dominant economic institutions at the global level parallels the emergence of environmental regimes outside these economic institutions. The issue-specific environmental treaties and regimes on (among other things) climate change, pollution of the oceans, transport of waste, acid rain, and biodiversity seem to move toward one another in terms of central policy principles, instruments, and legal basis, helped greatly by international conferences such as the 1992 UNCED in Rio de Janeiro. Developing countries outside the triad are connected with these emerging transnational economic and environmental institutions, agreements, and regimes in a different way than the countries that belong to the triad, and it is on that distinction that I will focus in this section.

Doubly Declining Environmental Sovereignty

National environmental regimes, policies, and programs in non-triad countries are under the influence of various "globalization" mechanisms that restrict the environmental sovereignty of these countries. This is to some extent true for all countries and a logical consequence of the globalization process itself, in which environmental deterioration and reform can no longer be monopolized by individual nation-states. However, for various reasons and by different mechanisms, the role of developing countries in setting up, designing, and implementing these transnational institutions and arrangements is rather marginal in comparison with that of the countries of the triad.

Spillover effects of domestic environmental regimes can be identified in several multilateral environmental agreements and international institutions. Specific environmental arrangements that have been introduced at the national level find their way to the international arena. And often it is the environmental practices and principles that have already been applied in triad countries that find their way into transnational institutions, owing to the more advanced environmental regimes and experiences of developed countries, their generally greater interests in the construction of international environmental regimes, and their more powerful position in international negotiations. The inequality in the knowledge and other resources brought to international negotiations have often been noted. Gupta (1997) has identified a number of reasons for this. A lack of manpower and detailed knowledge on international environmental negotiations, poor understanding of the country's own interests among its negotiator(s), difficulties in understanding the often complicated scientific models and methodologies used in environmental negotiations and agreements, and a lack of excellent national research organizations on these topics put developing countries at a disadvantage in assessing their own interests and the consequences of certain policy proposals for their home country, let alone proposing alternative models and methodologies that would serve their interests better. As a rule, developing countries can have a significant impact on negotiations only if they operate in large coalitions (e.g, the G77, OPEC, AOSIS) that are hard for the triad countries to ignore. Owing to conflicts of interest and the negotiating strategies employed by major triad countries, however, such coalitions are difficult to maintain.

The resources that developing countries have and use to strengthen their position in international negotiations on MEAs include natural resources and emission rights, in whatever form. The environmental resources are increasingly used to obtain financial compensation, major technology transfers, or other forms of environmental and economic assistance (e.g., the Ozone Regime and the Climate Change Regime). A creative coupling of distinct environmental regimes is coming about, with the developed countries using a carrot-and-stick strategy toward developing countries but also toward transitional economies. For instance, Global Environmental Facility funds were used as exchange value in order to get Russia comply with the Ozone Regime.

In some cases developing countries are not only deprived of their negotiating power owing to shortcomings in power and knowledge, but actually excluded. This occurred in the drafting process of the Multilateral Agreement on Investment, and also in the initial design of GATT, when most of the developing countries were still colonies of the triad nation-states. What we saw in the MAI in an exceptionally strong form happens in weaker forms in negotiations on most major multilateral environmental agreements: individual and small groups of developing countries, transitional states, and newly industrializing economies have little influence, even if they threaten not to sign a treaty. Even if they do not sign it, the developing countries and their domestic economic producers will in due course be forced to comply with it. In most environmental regimes, the developing countries are primarily "rule takers" and not so much "rule makers." International environmental standards are therefore increasingly perceived as protective measures of the triad to hinder local producers from entering the triad markets. Consequently, it should not surprise us to find Malaysia, India, and other East Asian states among the strongest supporters of initiatives to keep environmental standards out of international trade agreements. This was one of the central lines of argument of developing countries against the environmental and (child) labor protection ideas put forward during the 1999 Seattle summit of the WTO.

Developing countries and their industries are having a similar experience with the ISO 14000 series, a series of standards for environmental management developed in the framework of the International Organization for Standardization. The ISO is increasingly becoming recognized as the supportive technical authority facilitating the expanded role of WTO. Hence, ISO standards are becoming legitimate as tickets (or barriers) to market access, which might especially affect domestic developing country producers. As Di Chang-Xing (1999) has analyzed for China, governments and local industries do not have sufficient experience to implement corporate environmental management, and the cost of introducing such management is relatively high. Nevertheless, compliance with ISO 14000 standards is becoming an essential precondition for exports to the triad. Figures 7.1 and 7.2 illustrate December 1999 ISO 14000 registrations in selected countries around the world, showing the discrepancies between triad and developing countries. Meanwhile, developing countries (and environmental NGOs) have been only marginally involved in the drafting of the ISO 14000 series—and too late to make an effective contribution (Krut and Gleckman 1998).[13] All this makes it at least understandable if representatives of some developing countries interpret ISO 14000 standards in terms of protectionism and trade barriers, rather than as a fruitful basis for global industrial reform. And evidence from the sanitary and phytosanitary standards (figure 7.3) shows that such standards are indeed barriers.

As a consequence of this unequal loss of national sovereignty, the developing nation-states increasingly have to set their agendas, define their problems, select their priorities, and develop norms, standards, and governance models following the example of Western and international "standards," institutions, and practices. This not only reduces the sovereignty of developing countries in dealing with their own environment, it can also reduce the effectiveness of their national environmental regimes regarding existing local circumstances of pollution practices. The logic of technology transfer criticism regarding the "green revolution" indicates that the characteristics of the triad environmental regimes might not be "adapted" to

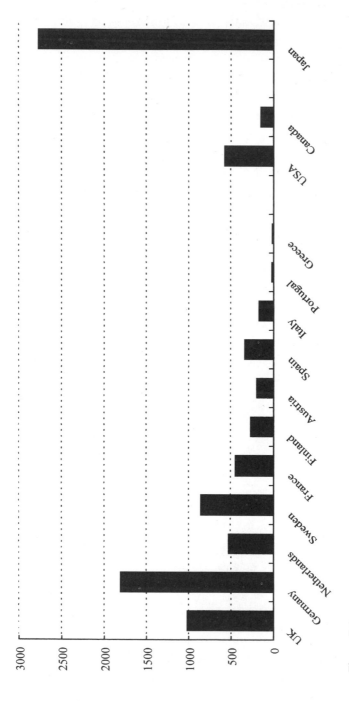

Figure 7.1
ISO 14000 registrations in triad countries in December 1999: EU, Northern America and Japan. Source: http://www.iso.ch.

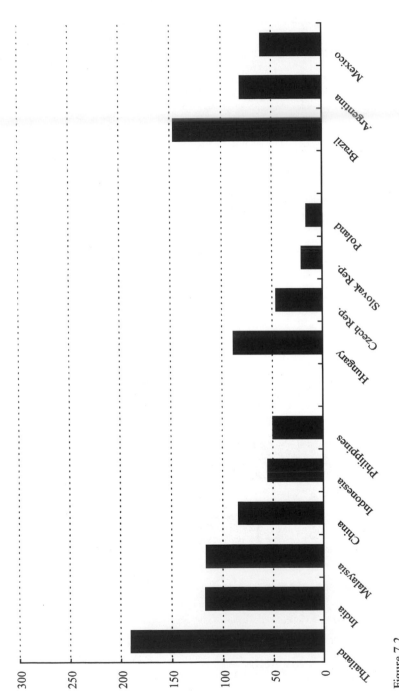

Figure 7.2
ISO 14000 registration in non-triad countries in December 1999: Asia, CEE countries and Latin America. Source: http://www.iso.ch.

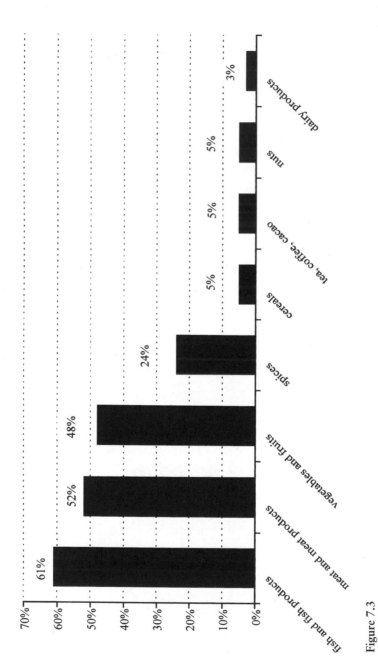

Figure 7.3
Percentage of developing countries whose agricultural and food products are banned from the EU market due to failures to fulfil SPS (Sanitary and Phytosanitary) Agreement standards. Source: Henson et al. 1999.

the local circumstances in developing regions. These more "practical" consequences are also reflected in the theoretical discussions regarding ecological modernization in an era of growing international interdependence and globalization, which to some extent brings us back to what was discussed in chapter 3. The crucial question in this debate is whether ecological modernization ideas and the forms of environmental policy making and reform based on them have any relevance for environmental reform in developing countries. This question becomes especially crucial in a situation where these ideas, principles, and practices are also at the foundation of the multilateral and supra-national environmental arrangements and institutions. There are very few scholars who would say that global environmental regimes should no longer have any impact on environmental reform in peripheral regions. Most would rather emphasize the need to harmonize environmental capacities and regimes up to at least the level that has been achieved in the triad countries.

National Environmental Regimes and Industrialization: Redistribution of Pollution

One of the liveliest debates on the relation between stringent environmental regimes in the triad and environmental policy and performance in the peripheral regions during the past 25 years has centered on the relocation of industrial pollution.

An introductory textbook on industrialization and the Third World (Chandra 1992) rightly states: ". . . most Third World countries . . . are committed to transforming or changing their rural-based agricultural economies to urban-based industrial ones. There may be differences in the level of industrialization they wish to achieve, the speed at which they wish to industrialize, or in their industrialization strategies, but nearly all of them are strongly committed to their goal of industrialization." Industrial investments in these countries are stimulated rather than discouraged, and foreign direct investments are often welcomed with favorable tax regimes, cheap facilities, and attractive sites. Sometimes lax environmental standards and regimes are also seen as a favorable condition for industrialization, especially regarding those industrial sectors for which environmental costs make a difference. Some believe that stringent environmental regimes in industrialized countries result in a relocation of especially heavily polluting or dirty industries[14] from these wealthier industrialized societies to peripheral or developing regions (LeQuesne 1996, p. 67ff). But in an era marked by globalization, is this an actual strategy on the part of polluting industries, a deliberate strategy on the part of nation-states, and/or a real tendency that can be empirically verified? Various studies have yielded such theses as the relocation of polluting industries, the "industrial flight" and "pollution haven" hypotheses, the migration of industries, and the reverse U-shaped relationship between industrial pollution and economic prosperity (better known as the environmental Kuznets curve). These studies all concentrate on the question of what happens with dirty industries and industrial pollution patterns when, in an increasingly globalized capitalist world, industrialized countries tighten their environmental regimes, production sites become less dependent on specific geographical locations, and developing countries are eager to attract foreign industrial investments.

Relocation of Pollution

In the 1970s—against the background of the first wave of environmental concern and the foundation of political institutions for environmental reform in most industrialized countries—Barry Castleman (1979a,b) was the first to investigate numerous examples of primarily US-based industries producing asbestos, benzidine dyes, and pesticides and processing arsenic and zinc that migrated to less developed countries.[15] Although the number of examples was impressive, the reasons behind these individual cases of relocation were not always obvious. It was not clear whether environmental considerations were important factors in the movement of these industries to less developed countries, or whether these individual cases caused a net increase of pollution in these developing countries (or in specific sectors in these countries) and a net decrease of pollution in the industrialized world. But in the case of asbestos industries relocating from the United States to Mexico, for instance, it became obvious that environmental and health considerations were among the main reasons to migrate.

In the 1980s, H. Jeffrey Leonard (1988) undertook an impressive study on the effects of increasingly stringent environmental policies in industrialized countries on the investment patterns of heavily polluting industries. Instead of using one of the various economic models of comparative advantage due to factor differences, Leonard developed and applied a more advanced social-political theory. His major finding—which is in line with the conclusions reached in similar studies by Levenstein and Eller (1980), Veeken (1982), Pearson (1987), Tobey (1990), Heerings (1993), Brown et al. (1993), and Zamparutti and Klavens (1993)—is that "the costs and logistics of complying with environmental regulations are not a decisive factor in most industrial decisions about desirable plant locations or in the international competitive picture of most major industries" (Leonard 1988, p. 231). Environmental considerations may initially make a difference (usually adding to already existing motives to migrate stemming from other factors), but changes in environmental regimes cannot be seen as strong enough to reshape the traditional economic and political forces that determine investment and location patterns. In addition, Leonard and others showed that under conditions of globalization and the formation of transnational regimes the possible economic gains in pollution havens usually last only a short time, that environmental protest and pressure are no longer limited to a few environmental front-runner countries but are spreading around the world through the activities of international environmental NGOs and global dissemination of information, and that governments are highly active in improving location conditions and offering compensation for specific disadvantages, so that it becomes inadequate to simply compare different locations on the basis of a cost-benefit analysis of the required environmental measures.[16] With the last argument Leonard and others have moved away from simple economic models in which the global distribution of capital investments is supposed to follow economic factors of production.

In the wake of the Brundtland report and the third wave of environmental concern, the early 1990s saw renewed attention to the migration of industrial pollution to less developed nations and to the relationship between industrial pollution and economic prosperity. However, the emphasis was more on the economic aspect and the balance of trade was now given a more central role. This economic emphasis was particularly strong in the rather simple economic model of Baumol and Oates (1988, pp. 258–265), which suggested that under conditions of free trade differences in national environmental regimes would affect the balance of trade and lead to "pollution havens." But numerous empirical studies conducted at the same time could not detect such effects. The economists Patrick Low and Alexander Yeats (1992) explored this problem in an indirect way by looking at the trade flows over time of products of "dirty" industries coming from different groups of nations. They found that over time the share of dirty products in the exports from less developed nations (with probably less stringent national environmental regimes) did indeed increase, and that it decreased in the exports from industrialized societies. As Low and Yeats explain, this trade pattern does not say anything about the actual changes in pollution by industries, since no information is available about pollution factors per production process and product (in material or GNP terms) in the various countries. Although Low and Yeats mentioned the possibility that differences in environmental policies could have led to this pattern, they listed many other possible and likely explanations. In line with a more restricted study on air pollution (Grossman and Krueger 1991), Hettige, Lucas, and Wheeler (1992; see also Lucas et al. 1992) found a displacement of toxic industrial production from the OECD countries in the 1960s to less developed countries in the 1980s, but this trend could not be related to more stringent environmental policies in OECD countries. Mani and Wheeler (1997) concluded that "pollution haven effects have not had major significance" and that wherever such effects occur "any tendency toward formation of a 'pollution haven' is self-limiting, because economic growth brings countervailing pressure to bear on polluters through increased regulation. In practice, 'pollution havens' have been as transient as "low wage havens.'" The German political scientist Martin Jänicke (1995) similarly analyzed the trade balances of OECD countries and developing countries for several polluting industrial sectors (fertilizers, chemicals, petroleum, paper and wood pulp, non-ferrous metals) and concluded that the general trend is not so much that of a relocation of "dirty" basic industries to developing countries but rather a production surplus of these "dirty" sectors in the industrialized world.[17] The reasons for this, according to Jänicke, are the relatively minor costs of pollution control and the capital-intensive and knowhow-intensive character of these "dirty" industries (chemicals, etc.), favoring a location in developed, industrialized nation-states. Of course, by looking at industrial sectors and not analyzing more disaggregated levels, Jänicke and the other authors quoted were not able to identify possible differences within one sector or product category (between, for instance, different pesticides or different production processes of similar products). Interestingly, Jänicke also carried out an analysis comparing alternative strategies that were assumed to lead to the ecological improvement of "dirty" industrial sectors in triad countries. He found that ecological modernization—i.e., the ecologically more rational and efficient (re)use of material, energy, and natural resources—is the basic strategy for environmental improvement of dirty industries in OECD countries, and not migration. Case studies of developing countries, such as J. Timmons Roberts's (1998b) study on the chemical industry in Brazil, also suggest that globalization and global environmental standardization spell the end for any "pollution haven" advantage. Fear of litigation, negative public relations, and concern about a loss of triad markets led many transnational and export-oriented chemical firms in Brazil to adopt global standards (e.g., ISO 14000), thereby improving some of the worst cases of environmental performance even in remote areas that are poorly "visited" by the increasingly powerful environmental movement.

The Environmental Kuznets Curve

In some ways, the opposite of the "pollution haven" and "relocation" arguments is the idea of an environmental or "green" Kuznets curve. Although "relocation" or "pollution haven" ideas would indicate that movement of capital from North to South is detrimental to the environment, the green Kuznets curve indicates that more capital and economic development going to the South is only better for the environment. A large number of empirical studies and a major debate have emerged in the wake of the World Bank report (1992) in which this "environmental Kuznets curve" was cited approvingly. The inverted U curve suggests that a universal relationship exists between economic development (in GNP per capita) and environmental pollution intensity per unit GNP (especially air and water quality). Although the poorest countries have a relatively low rate of environmental pollution, this is augmented up to a certain maximum with increasing economic development. The industrialized OECD countries, in turn, would show a drop in the environmental burden per unit GNP (figure 7.4). Several authors have—with good arguments—questioned the empirical validity of this curve and the extrapolation that implies that the same process will occur in all cases and countries. They have pointed at the selective use of environmental indicators (only some emissions and no resource stocks are involved in the empirical studies), limited and unreliable data, contradictions between studies, and a bias toward "best fit" curves (Arrow et al. 1995; Ekins 1995; Wallace 1996; Stern, Common, and Barbier 1996).[18] Moreover, the policy conclusion that economic growth is a panacea for improving environmental quality is

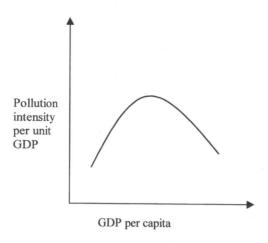

Figure 7.4
The Environmental or Green Kuznetz curve.

severely criticized from various sides. Some of the critics even suggest that a N curve rather than an inverted U curve can be analyzed, suggesting that any improvement is in the end "neutralized" via economic growth—an idea already launched in almost exactly the same terminology for industrial economies by Martin Jänicke (1978) (figure 7.5).

The World Bank has been advocating and legitimizing another element of neo-liberal policies with an appeal to environmental claims. The growing consensus that balance-of-trade and pollution-haven effects are not substantiated by empirical evidence led the World Bank to publish a number of studies (Low 1992; World Bank 1993) concluding that countries pursuing more open trade policies tended to have lower growth rates in the pollution intensity of production (measured per unit GNP) than countries pursuing import-substitution industrial policies. This suggests an environmental legitimization for the structural adjustment policies (open markets, diminishing trade barriers) that were so strongly pushed by the World Bank and the International Monetary Fund from the mid 1980s on. These World Bank conclusions on the relationship between open trade policies and superior environmental performance have come under serious criticism. Rock (1996b), for instance, provides empirical data that show the opposite: over 1973–1985, countries with export-oriented trade policies had higher pollution intensities of GDP than countries with inward-oriented trade policies. Glover (1995) argues that structural adjustment programs can be either

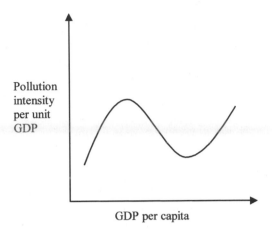

Figure 7.5
The N-curve. Source: Jänicke 1978.

benign or malign for the environment, depending very much on how these programs are designed and implemented.

Conclusions and Consequences

The general conclusion of these only partly comparable studies dating back to the late 1970s is that the increasingly stringent environmental regimes of the industrialized nation-states and the transnational institutions dominated by these states have not resulted in a massive migration of dirty industries to the Third World. The environment is not considered an important location factor by investors, and the World Bank's 49 "competitiveness indicators" do not even include environmental regulation (Gentry 1999). There is evidence of a limited number of cases in which relocation of industries and investments from industrialized countries to developing countries was primarily induced by differences in environmental regimes.[19] There is also evidence of occasions on which other location factors caused industries to migrate to non-triad countries, with a net increase of pollution occurring as a side effect (e.g., some Japanese investment moving to the countries in the region; see chapter 6). Industries that migrate because of inferior product quality, poor labor conditions, or illegal economic practices may not always be the cleanest producers. But generally, the development of more stringent environmental reform in industrialized countries appears to result in movement toward ecological modernization of industrial production in these industrialized countries (i.e., more environmentally rational and efficient production) rather than migration.

This is not the end of the story, of course. For one thing, environmental regimes may become more important location factors if environmental considerations gain in political and economic importance in some industrialized countries, while environmental interests in some non-OECD countries stay far behind. In a globalizing world where international coordination of national environmental measures and policies has become a growing concern, national environmental regimes tend to converge rather than diverge. Although major differences are not impossible, increasing importance of national environmental regimes as location factors will most probably be accompanied by increasing international demands and by pressure to harmonize them. That is in fact what we have been witnessing in the EU and NAFTA regions, and globally with respect to the debate on environmental provisions in the WTO/GATT. Furthermore, the fact that empirical evidence of environment-induced migration of ecological risks via industrial relocation is poor does not signify that triad countries do not capitalize on the environment of non-triad countries. All kinds of more indirect and less intentional mechanisms can and do result in a sometimes increasingly unequal consumption of the world's natural resources by triad countries, or regions within the triad countries, relative to the non-triad countries. Or, as Zarsky (1999, p. 3) concludes,

While "pollution havens" cannot be proven, a pattern of agglomeration of pollution is discernible, one based not on differences in national environmental standards, but on differences in income and/or education of local communities. They may not be "havens," but there are clearly "pollution zones" of poorer people, both within and across countries, where firms perform worse and where regulation is less effective.

During international negotiations such as those within the Framework Convention on Climate Change, scholars, and political representatives of developing states have emphasized this unequal capitalization of the world environment (Sachs 1993; Redclift 1996; Sachs et al. 1998). Although it does not make this unequal allocation of environmental space less troublesome, such consequences are generally not caused by stringent environmental reform in the industrialized countries. Consequently, neither demands for laxer environmental regimes in triad countries, nor calls for more stringent regimes in peripheral regions can be based on these kinds of arguments. Finally, some scholars have argued for the opposite, and the leading institutions of the global economy (the WTO, the IMF, the World Bank) as well as neo-liberal free-market ideologists have been the first to embrace those ideas, to the effect that foreign direct investments lead to improved environmental performance in developing countries. On a national level, these ideas have been advocated by the World Bank, among others, and I have weighed them above. On a business level, firms from triad countries investing in developing countries are believed to have significantly better environmental performance, creating "pollution halos." In its initial writings on globalization and the environment, the OECD— and especially its Business and Industry Advisory Committee—developed its arguments in that direction (OECD 1997a,b). Analyzing the statistical evidence, Zarsky (1999) concludes that such a pattern does not exists once firm size is taken into account. In some sectors and some countries, large foreign firms in particular have better technologies and closer links to green or environmentally sensitive consumer markets. But in general, foreign links, including export markets and ownership of plants, do not seem to make much of a difference.

Epilogue

In the age of globalization, all nation-states—not only those belonging to the triad—have become more and more interdependent. It should not surprise us, therefore, if we find this also to be true with regard to "environmental sovereignty" (Mol and Liefferink 1993) and environmental reform. On the premise that this environmental interdependence is emerging in an unequal world, with triad countries dominating the economic, political, cultural, and environmental agendas, stringent international and national environmental regimes are often said to create negative (environmental and economic) consequences for developing countries. As a result, non-triad countries use their environmental capital in international negotiations to obtain economic and technological compensation while trying to block stringent environmental policies that might jeopardize their export position and their competitive position in the global economy. However, environmental arguments are used just as often, and by various national and international players, to strengthen particular economic perspectives or ideologies.

Industrializing Economies in a Global Environment

As I illustrated and argued in the previous chapters, developing countries are often believed to be the victims, also the environmental victims, of globalization. Although it may be too simple to draw a direct relationship between the strictness and effectiveness of national environmental policies and the per capita national income or gross domestic product (as Dasgupta et al. (1995) suggest), it is nevertheless evident that developing countries generally have fewer resources to establish stringent national environmental regimes to cope with the environmental side effects of globalization. By the same token, the environment in these developing countries might be the first to benefit from environmental reforms induced by globalization, exactly because their national capacity for environmental reform falls short. In chapter 7 I showed that developing nation-states generally do not have equal access to the design processes of transnational institutions for environmental reform. This chapter will concentrate fully on environmental reform within countries that are not part of the economic triad. How do these countries experience globalization trends? How, and to what extent, do these globalization processes interfere with the domestic attempts or failures to make an attempt at environmental protection and reform? By which national institutions do globalization dynamics interfere in the environmental performance and regimes of peripheral nation-states? Can distinct categories (areas, sectors, etc.) be identified within a specific country that show different ways of coping with the environmental consequences of globalization processes?

It is beyond the scope of this volume to provide an extensive analysis of a representative sample of countries from the "South." Besides, such an analysis is not required for my purpose, which is to identify some of the main processes and mechanisms of environmental deterioration and reform at work in developing countries under conditions of globalization. In view of this goal and the research programs in which I have been involved,[1] I will look at three countries that can to some extent be taken as examples of larger groups of states with similar characteristics. Vietnam is a typical example of a second-generation newly industrializing economy[2] with some typical transitional characteristics. Kenya is an example of a sub-Saharan African developing country with a predatory state[3] whose industrialization appears to stagnate. Curaçao is a representative of so-called small island development states, which are often believed to be among the first environmental victims of climate change and other global environmental problems while hardly contributing to them.

Each of my three case studies will begin with an outline of the country's environmental profile, both in terms of its main environmental polluters and in terms of the capacity of the state to deal with them. Subsequently, I will concentrate on how these local or national dynamics in environmental pollution and reform intersect with globalization processes and dynamics, with special attention for the value of the ecological modernization type of reforms I have hypothesized theoretically.

A Second-Tier Newly Industrializing Economy: Vietnam

If we want to analyze the present and future environmental reform possibilities of Vietnam's industrialization process, we should not only be aware of the typical characteristics of its contemporary economic and environmental developments; at the same time, we should realize that Vietnam is not "just another developing country" (if such a country exists at all). Vietnam can be "conceptualized" at the crossroads of two processes: the transformation of a command economy to a market-oriented growth model (a process also seen in other transitional economies) and the more traditional transition of a less developed country to a newly industrializing economy (a process also seen in an increasing number of East and Southeast Asian states). Only China seems to have a similar position at this crossroads, although the size of the country, the economy, and the scale of environmental deterioration are of a different magnitude.

Between a "Developmental State" and a "Transitional Economy"
Like other newly industrializing countries in East and Southeast Asia, Vietnam shows a rapid process of industrialization. In that respect Vietnam

is a typical example of an East Asian "tiger" economy, although its economic development began later and is still lagging behind countries such as Taiwan, South Korea, Singapore, Thailand, and Malaysia. Vietnam can be understood as a country governed by a kind of so-called developmental state (Evans 1995; Castells 1997), similar to other East and Southeast Asian "tiger" economies. The Vietnamese state has a tradition of strong involvement in economic development, embedded in the economic and social networks surrounding the state. In economic terms, Vietnam is currently undergoing a process of rapid economic and industrial growth that began in the early 1990s, not unlike the developments in some of the first-generation Asian "tiger" economies that began some 20 years ago. At the moment, Vietnam differs from the "real tigers" in that it is a "newly industrializing periphery" rather than a high-performance economy (Rock 1996a). What makes Vietnam special is the relationship between this rapid industrialization and economic development and the change in the economic system that is taking place at the same time: the transition from a pure command economy toward a market-oriented growth model. Sikor and O'Rourke (1996) provide a summary of the institutional transformations framing contemporary economic activities, following the economic reforms in Vietnam since 1986: the government has liberalized economic production and exchange; resource allocation has shifted toward market mechanisms, with the goal of increasing flexibility and efficiency; owing to state enterprise reform, the 1993 Land Law, and tax reforms, assets have been transferred to the private sector, strengthening its role; and international trade and investment have been liberalized by the 1987 Foreign Investment Law and more recent foreign trade reforms. This means that the integration of Vietnam into the world market and its experience of environmental and other consequences of globalization trends are relatively recent and, up to now, only partial.

The institutional characteristics and transformations taking place in Vietnam would lead one to expect to see some similarities between Vietnam and other transitional economies, for instance those in Central and Eastern Europe. But there are at least four major differences that prevent us from drawing too close parallels between these countries:

• The centrally planned economies of Europe were on a different course in terms of economic development and industrialization before their shift to a market-oriented growth model.

• Vietnam's economic transformation seems to proceed more slowly and better "controlled" than the shock therapies that most European transitional economies are going through.

• While in Europe economic transitions were in most countries accompanied—or even preceded—by democratization, this is less so in Vietnam.

• In contrast to Central and Eastern European economies, Vietnam had little if anything in the way of a system of environmental policy making until 1986 and such a system had to be created in the early days of economic reform. In a way, this saved Vietnam the trouble of having to go through a difficult reform process of existing but inadequate and old-fashioned environmental institutions and legal systems, a process that is so characteristic of contemporary transitional societies in Central and Eastern Europe. The lack of such additional complications provides Vietnam with at least a theoretical advantage of being able to directly develop adequate environmental institutions directed at preventing and neutralizing the harmful environmental consequences of a market-oriented industrialization process in a globalizing world. In practice, however, the advantage is not so obvious.

Industrial Development and Environmental Devastation

Since the economic reform programs of the mid 1980s (the so-called *Doi Moi*), Vietnam has first shown a period of poor economic performance, with high inflation and limited economic growth. Since 1990 however, Vietnam is showing an average annual economic growth of 8–9 percent (ADB 1998), slowing down by the turn of the century to some 4 percent. Inflation has fallen below 3 percent in 1997. The 1998 Asian economic crisis seems to have affected the Vietnamese economy only to a limited extent as compared with its neighbors, although exports dropped somewhat and foreign direct investment (especially by East Asian companies) stagnated. This rather modest influence of the Asian crisis on the Vietnamese economy should be attributed to the fact that as yet the country is still little integrated into the world market and therefore less dependent on it, although with FDI mainly coming from ASEAN countries Vietnam is rather vulnerable to regional crises. The recent decrease in economic growth and FDI figures is often explained by the continued heavy political control of the market and the disappointing speed of liberalization.

Agriculture and forestry have traditionally been the major contributors to national income, providing around 40 percent of GDP, although the 1997 World Development Report uses 28 percent and the Asian Development Bank (1998) estimates it at 31 percent. Agriculture and forestry are also the main sources of employment, with some 70 percent of the labor force. Only about 13 percent of the total employed labor force work in the industrial sector (data on 1990–1993). However, the economic importance of the industrial and service sectors is on the rise as compared with agriculture and forestry. Industry accounted for 25–30 percent of GDP in 1996–1997 (Mol and Frijns 1998; World Development Report 1997; ADB 1998) and its share is expected to rise to 35 percent in 2010 according to projections of the Ministry of Planning and Investment. The industrial sector consists of national state-owned enterprises (SOEs), provincial state-owned enterprises, and the non-state sector made up of cooperatives and private companies. The emphasis in industrial development has traditionally, before *Doi Moi*, been on the heavy industry. Nowadays, with the increasing importance of a market economy system, light industry, with consumer products such as textiles and food products, is becoming dominant. A substantial part of the industrial development is taking place in newly established industrial zones in the Ho Chi Minh City and Ha Noi regions and Dong Nai Province.

From 1993 up to the 1998 Asian crisis, industrial output has grown by 13–14 percent annually (ADB 1998).[4] The number of private firms in the formal industrial sector has grown, but most are still small. In that sense, the transitional process of Vietnam clearly differs from the shock therapies in Central and East European states. Vietnam is only very cautiously treading the path of privatization, to some extent successfully modernizing the SOE sector by acquiring new technology through joint ventures between state-owned enterprises and foreign capital. Or, as Jansen (1997) puts it, there has been a transition to the market, but only a limited transition to a private sector economy. Apparently, and partly owing to government policy, foreign investors also prefer joint ventures with SOEs rather than private firms, largely because the SOEs have privileged access to land and credit (Irvin 1996). Foreign direct investment has grown steadily since the Foreign Investment Law of 1987, up to some US $19.4 billion in 1995, of which over half goes to the oil and gas sector and 10 percent to the industry sector (UNDP 1995).[5] The oil and natural gas industry is dominated by foreign capital, as is food processing and light manufacture.

Doi Moi has liberated market forces that have increasingly directed investments into sectors where Vietnam enjoys a strong comparative advantage. These include both labor-intensive industries (such as textiles, clothing and footwear) and resource-intensive industries (such as fossil fuels, plantations, agricultural processing and building materials). Unfortunately, some of these sectors are by nature not particularly environmentally sound. For example, petroleum refining and the production of petrochemicals can be accompanied by serious pollution problems. Agriculture and food processing are major sources of organic pollution (water and waste), and textile dying, electroplating, leather tanning, and pulp and paper production generate significant amounts of organic and hazardous wastes.

It was estimated that in the mid 1990s more than 3000 major industrial enterprises in Vietnam were discharging wastewater without any treatment. As a result, in the major industrial centers, dissolved oxygen in surface water is always near zero, BOD and COD are high, and metals and other toxic substances are regularly found in high concentrations. Urban air quality is also deteriorating, principally owing to the proliferation of motorized transportation that come with the intensification of urbanization and industrialization. Ambient concentrations of pollutants such as carbon monoxide, nitrous oxides, sulfur oxides, lead, and particulates exceed emission standards in many areas. In the field of solid waste management, the majority of waste is generated by domestic sources, and only 18 percent is of industrial origin. There is no systematic monitoring system that can provide detailed and reliable data on the amounts, the contents, the origins, or the destinations of different types of industrial solid waste. Neither is there any form or system of hazardous waste management. In short, the rapid industrialization is not being accompanied by measures to control pollution and this has resulted in an increase of wastewater discharge, air pollution, energy and natural resource consumption, and release of industrial solid waste.

Government Response to Environmental Change

In the early 1990s, Vietnam was only beginning to construct its institutional framework for environmental management and policy. Like in most developing countries, environmental policy in Vietnam still follows a pretty much conventional model of command and control that is characterized by laws, standards and regulations as the main instruments, and top-down implementation of legislation. Not unlike the policy models in Western countries in the early 1970s, this policy style can be ineffective, bureaucratic, falling short on implementation and enforcement, and not accruing technological innovations. As I have argued before (Mol and Frijns 1998; see also Frijns et al. 2000), it should be concluded that this command-and-control approach has had only limited success in Vietnam until now. First, compliance with environmental regulations can only be enforced under threat of fines or punishment; however, for cultural reasons, courts are only used as a last resort in many developing societies, especially in Asia, which means they are rarely used. Second, detailed and extensive environmental regulation cannot be managed effectively given the contemporary budgetary, manpower and administrative constraints. Third, a low degree of devolution of authority to local governmental agencies reduces the country's capacity for monitoring and enforcement. In fact, neither the Ministry for the Environment nor the local environmental authorities can effectively impose penalties on violators. Particularly difficult to control is the pollution caused by large state enterprises. Local enforcement agencies often feel reluctant to exercise their power on these enterprises. Fourth, fines are usually set too low to deter violators, they remain unchanged in nominal terms for years, and become eroded by inflation. The rent-seeking behavior of the enforcement officials is another factor reducing the effectiveness of government control.

Although major improvements have been made on paper and are under construction in the institutional layout and the development of environmental laws and policy instruments, implementation until now has been poor (Cao Van Sung 1995; O'Rourke 2000). This should not surprise us much, since this seems to be the rule than the exception in most countries that face both rapid industrialization and rapid development of their institutional framework. A coherent institutional framework and legal rules are essential preconditions for the construction of an effective system of industrial environmental management, therefore it is not illogical for states to give some priority to legislation and the construction of an institutional framework in an initial phase of environmental reform. However, for this institutional structure under construction to mature and be consolidated, successes in implementation are essential. This means that more attention will most likely be devoted to implementation and enforcement processes in the years to come, in order to turn the institutional and legal progress into actual environmental improvements.

Global Determinants of Environmental Deterioration and Reform

Although in the triad countries the national environmental regimes came about rather independently from global developments in the 1970s, contemporary national developments are highly dependent on transnational and global processes. This is one of the reasons why one cannot expect the construction of an environmental regime to follow the same course in Vietnam as it did in the triad countries in the 1970s and the 1980s. The question then becomes whether transnational and globalization processes will lead to a stagnation of ecological modernization in countries such as Vietnam? Or could globalization processes instead serve to trigger or accelerate environmental reform processes in Vietnam?

It does not require great analytical skill or imagination to conclude that the opening up of Vietnam's economy, starting from 1986, has greatly furthered its integration in a globalized world economy. The increase in foreign direct investment in industry, the rapid growth of industrial exports, the diversification of the industrial structure, and the growing dependence of Vietnam's economic performance on external economic developments and actors all point in that direction. The transitional economy in Vietnam shows some major differences with those in Central and East European countries in terms of, among others, the degree of privatization, the extent of state control on the economy, the political changes that parallel the economic transformations, the influence of foreign capital on local markets. Nevertheless, Vietnam is fast on the way of becoming integrated in the world market. It goes without saying that this transformation of the economic structure is reflected in changing patterns of environmental deterioration. To some extent the changes are related to the flow of "additions and withdrawals," the use of natural resources and the growing emissions into the environment. The scale of industrial production and the concentration of production activities in specific regions add to the environmental problems that are particularly grave in the three major industrial areas, but not limited to those areas alone. One at least equally important consequence of Vietnam's ongoing integration in the world economy is related to the quality of environmental transformations, as new industries emerge and formerly unknown substances, products and production activities now affect the country's environment. This does not always have to result in an aggravation of environmental problems, however. As Reed (1997, p. 270) argues, a diversification of the economy away from the resource sectors such as forests, fisheries and land can lead to a more sustainable development, though not automatically. Lastly, it should be mentioned that Vietnam is actually losing control over its industrial production. It is not just that the transformation from a centrally planned to a market economy has reduced the government's control on investment, strategic developments, or siting. As foreign direct investments become more important, either entirely foreign or in the form of joint ventures with state or private enterprises, national control over the environment is reduced. Although it is not predominantly heavy industries such as steel or chemical bulk production that have moved into a liberalizing Vietnam, the oil and gas sector, food production, and the paper and pulp industry—to name but a few sectors—are all polluting sectors that have accelerated rapidly with foreign investments. In addition, industrial recycling is affected. Some recycled products are struggling in the competition with better quality (perceived or real) products made of virgin material by foreign firms in or outside Vietnam. In other cases, Vietnamese industries have experienced a drop in demand for their solid "waste" to be used by domestic recycling industries, since the latter prefer the large amounts of more uniform "wastes" imported from elsewhere in the region, which in turn has resulted in increased dumping of industrial solid waste (Nguyen Phuc Quoc 1999).

As discussed above, the rapidly expanding institutional frameworks for the environment are as yet far from able to cope with this accelerating economic liberalization and internationalization. This is not only because of a lack of personal and institutional knowhow as regards the environment, it is also caused by a growing concern in Vietnam with international competition for capital. Even though government coordination seems stronger than in most of the other East and Southeast Asian "tiger" economies, struggles and conflicts of interest between governmental agencies do exist, and international competitiveness is increasingly used as an argument in these domestic struggles. At the same time, the opening up of Vietnam and its integration into the globalized world has triggered or stimulated initiatives to counteract the growing environmental disruption, for instance in order to comply with MEAs (such as the Montreal protocol). International donor agencies and institutions such as the United Nations Development Program, the United Nations Industrial Development Organization, the World Bank, the Global Environmental Facility, the United Nations Environment Program, the Asian Development Bank, and bilateral support programs of Canada, Sweden, Australia, Japan, the Netherlands, Germany, Switzerland, and other countries support the technical, institutional, and human resource capacity building of governmental institutions working on the environment (Frijns et al. 2000). Until now, regional institutions such as APEC and ASEAN have not seemed to be of particular importance in building national environmental infrastructures. Although Vietnam's environmental sovereignty is sometimes jeopardized significantly, the effectiveness and sustainability of these projects and programs can be criticized, and major parts of this aid flow back to the donors via consultancies and environmental investments, the net effect is nevertheless positive: Vietnam's capacity to develop environmental policies, monitor environmental deterioration and implement measures has increased, be it not to the level that is necessary to control and redirect the rapidly growing foreign industrial investments.

Foreign industrial investments have also introduced positive improvements in the environmental reform systems in Vietnam. International Organization for Standardization guidelines were first introduced via foreign investments. Wastewater treatment and other environmental facilities are most commonly found in foreign-owned enterprises or joint ventures rather than in Vietnamese enterprises. The system of environmental impact assessments—arguably the most promising system of environmental reform that Vietnam has (Mol and Frijns 1998)—works best at major foreign enterprises, and the environmental expertise available at joint ventures or foreign-owned companies is usually easily mobilized to review environmental performance and improve production. On the whole, foreign companies are not among the most harmful companies, although examples of Nike (poor labor conditions), Dona Bochnag Textiles (air pollution), and the Vedan tapioca factory (surface water pollution) would appear to prove otherwise. In addition, domestic firms that produce for export markets outside the region sometimes improve their environmental performance in parallel with their improvement of product standards (necessary to enter these markets), proving that product quality and environmental quality are often related. But in contrast to what Vogel (1997, p. 360) has found for industrialized societies, stricter domestic environmental standards have not turned out to give domestic firms in Vietnam a competitive advantage.

In addition, the opening up of Vietnam has led a steady flow of international NGOs to open up local offices in Vietnam. While the transitional economies in Central and Eastern Europe saw a rapid increase of especially local and national environmental NGOs, Vietnam has until now severely frustrated the setting up of national environmental NGOs (Eccleston and Potter 1996; Mol and Frijns 1998). An active environmental civil society is largely nonexistent. Only some local communities directly confronted with the negative environmental effects of industrial production occasionally manage to deflect investments into more sustainable directions (O'Rourke 1997 and 2000).[6] Vietnam only allows foreign NGOs, be it with certain restrictions. The World Wildlife Fund, the International Union for the Conservation of Nature, and Environment and Development Action are among the best-known international environmental NGOs that have recently started operations in Vietnam. Their tactics are not as confrontational as those of similar organizations in other more industrialized countries in the region or beyond. They work rather closely together with governmental agencies and neither oppose the government (as in European countries or the United States), nor leave them aside to take over governmental tasks (as we see in some African and Asian countries). This seems to fit rather well with the characteristics of a developmental state. And even if the international NGOs have not had more than rather meager influence, we can identify a rapid emergence of international environmental ideas and ideologies in the various sections of Vietnamese society, from the state organizations to educational and research institutions. There is little doubt that the step toward national and local environmental NGOs will be taken in the near future.

It is of course impossible to predict the outcome of these diverse and sometimes contradictory globalization forces for environmental deterioration and reform in Vietnam. It depends to a major extent on the way the Vietnamese government will deal with these global interventions and interferences locally, and to what extent it wishes and manages to locally strengthen those global processes that enhance national environmental reform, while checking and counteracting the more environmentally harmful ones. The fact that the country still has a strong state structure—partly a leftover from its centrally planned economy period, partly characteristic for the region—can be an asset for environmental reform, especially compared to our other two examples, Curaçao and Kenya.

Comparing Vietnam and China

The economic, political, and environmental importance of China in the world makes it worthwhile to go a little deeper into the similarities and differences in globalization-environment relations between China and Vietnam.

The general picture of Chinese national environmental policy does not diverge fundamentally from that of Vietnam, but there are differences. China has been building its state environmental protection system for a longer time than Vietnam and is more advanced and to some extent also more effective (Rock 2000). In general, a centralized system has more difficulties in operating effectively in a large country such as China than in smaller countries. Although we can witness the various coordination problems between the various Beijing-based ministries, as well as between the Beijing-based State Environmental Protection Agency (SEPA) and the decentralized environmental authorities in provinces and cities (Jahiel 1998; Vermeer 1998; Rock 2000), the relative autonomy and radical measures seen in local environmental policies in the late 1990s are remarkable. Shutting down factories for environmental reasons no longer appears to be an exception in China, while it is certainly not the rule in Vietnam. The fact that an organized national environmental movement is almost absent (Qing and Vermeer 1999) and the international environmental NGOs face considerable difficulties in operating in China (although nature protection organizations such as the WWF have strong programs) implies that systems of community complaints and media coverage are crucial signs for environmental authorities to act (Dasgupta and Wheeler 1996; Wang 2000).[7] The steady economic growth percentages (especially in the East and the South), the rapid industrialization, the growing urbanization and the absolute number of consumers and producers place Chinese environmental policy before a challenge that is unprecedented in history. Clearer water and bluer skies (World Bank 1997b) are not easy to attain. How do globalization processes interfere with that?

By the late 1990s, around 15 percent of total domestic investment in China was FDI, 80–90 percent of which was invested in manufacturing, real estate and construction. FDI contributes to around 5 percent of GDP and 10 percent of manufacturing value added, according to World Bank figures, which is roughly comparable to Vietnam.[8] The majority of FDI comes from triad TNCs, while the percentage of FDI from overseas Chinese has been diminishing recently (Ramstetter 1998). Like in Vietnam, in China these foreign investors are not among the most polluting industries either. If we can take ISO 14000 standards as an indicator, over 70 percent of the 232 certified companies in China (by the end of 1999) were joint ventures or foreign companies. Di (1999) carefully analyzes the difficulties that

Chinese companies experience in meeting the ISO 14000 standards as compared to triad companies, and which explain these figures. For a long time the Chinese home market was vast and closed, so that domestic companies did not have to worry too much about international markets, competition and trade, or international (environmental) standards. This is rapidly changing as the market is opening up and China is soon to become a member of the WTO. Especially larger Chinese firms operating on the international market are rapidly catching up with modern production and management methods, to some extent also including international environmental standards. On the other hand, the Chinese authorities are hesitant in international negotiations to support stringent environmental policies that could rebound on domestic efforts. (See Johnston 1998 and several other contributions in McElroy 1998.) The deadlock between the United States on the one hand and the non-annex I countries China and India on the other with regard to the Kyoto protocol on greenhouse gas reductions shows this clearly. China refuses to make an effort to reduce greenhouse gases, while the United States refuses to sign and ratify the Kyoto protocol if the large non-annex I countries remain unwilling take some measures. Similar processes can be expected now that China is to develop from a "rule taker" into a "rule maker" in the WTO, sharpening the environment-trade and green protectionism controversy between North and South (chapter 7 above). As China is a major player in the world economy and polity, its position cannot be taken as lightly as the interests of Vietnam.

At the same time, the potential transboundary environmental consequences of an unsustainable Chinese industrialization path are reflected in the priorities of international environmental aid toward China. Asuka-Zhang (1999) gives a clear example of environmental Official Development Assistance (ODA) and environmental technology transfer from Japan to China, and the problems of conflicting priorities and interests between the donor countries and actors and the Chinese authorities. Equally, China has recently become an object of considerable international attention as well as environmental funding, via several MEAs and multilateral institutions (Huq et al. 1999).[9] The phaseout of chlorofluorocarbons after the Montreal protocol has been such an example. In the first few years after the Montreal protocol negotiations (1987) China increased its chlorofluorocarbon production (by some 100 percent between 1986 and 1994; Held et al. 1999, p. 397), becoming the world leader in chlorofluorocarbon production and consumption in 1996. It then moved to stabilization and even decline in some sectors (such as refrigeration), in response to international aid and potential trade bans by triad countries. The growing Chinese share of world carbon dioxide emissions is very far from any such control, as primary commercial energy consumption is still increasing by more than 5 percent annually, and energy efficiency is still low.

Although the international and global mechanisms stimulating environmental reform in China show clear similarities with those in Vietnam, the mere size of the economy and the (geo)political importance of the country do make a difference.

"Small Island Development States": Curaçao and the Netherlands Antilles

Whereas Vietnam stands as a model for the combination of a developmental state with a transitional economy, Curaçao and the Netherlands Antilles bear the typical characteristics of what have become known as small island development states. Before providing some information on the political and economic structure of Curaçao (the largest island of the Netherlands Antilles) and analyzing the way globalization tendencies have affected—and restore—the environment on Curaçao, I will outline some of the central features of small island development states.

Politics, Economics, and Environments of Micro-States

Micro-states, and particularly small island development states, are usually considered to have two central characteristics: their small[10] size (whether measured in inhabitants, square kilometers, or economic criteria such as GDP; see Bray and Parker 1993) and their geographical insularity. The latter characteristic is usually reinforced by a more socio-cultural insularity following a specific historical development. The contemporary language, culture, religion, institutional design, and economic and trade relations are strongly shaped by early colonial times, preventing closer collaboration between neighboring micro-states.

In contrast to Vietnam and other transitional economies, the economies of micro-states have been open for quite some time, hence dependent on world economic developments. The limited home market rules out cost-effective mass production without major exports. Most micro-states, but especially the developing ones, have greater difficulties in diversifying their economies than larger states, and as a result become heavily dependent on import (Streeten 1993; Briguglio 1995). These countries usually have a comparative advantage for certain types of products, which contribute to the major share of exports. Owing to the limited geographic scale of most small islands, agriculture is of limited importance or restricted to one or two crops (coffee, sugar cane). In addition, most micro-states export the major share of their products to a limited number of states, which adds to their dependence on developments in these states and limits their possibilities to design their own macro-economic policies, a situation which has been described as "insular policy impotence" (Villamil 1971). This dependence is not only restricted to a limited number of states but often also to a limited number of transnational corporations, which occupy a crucial—if not to say monopoly—position regarding exports or export-oriented production in micro-states. One consequence is that trade is vital in GDP. In large economies trade accounts for some 15 percent of GDP. In micro-states this can easily run up to 50 percent, again increasing their dependency on world economic developments.

Diseconomies of scale are not restricted to the economic sector of industrial and agricultural production and marketing. The provision of social and physical infrastructures and adequate governance are also confronted with disadvantages related to scale. The cost per capita of sewerage, drinking water, electricity, and transport are on average higher within micro-states, despite the often shorter physical distances between locations. The same is true for health care, education, and environmental management, to name but a few aspects. In these sectors, specialization is possible only to a limited extent owing to cost factors. In addition, most small development states have inherited a large bureaucracy, with all the attendant financial and other problems. Reductions of state bureaucracies seem especially difficult in relatively closed communities that are characterized by specific configurations of social relations. Although cooperation between micro-states would be the logical answer to these diseconomies of scale, and although some successful examples do exist (CARICOM, the University of the West Indies), most small island development states are still narrowly focused on the larger states in the region or the former colonizers, for various kinds of reasons.

These economic and political difficulties notwithstanding, small states also profit from some comparative advantages that are illustrated by the economic success stories of some of them, such as Singapore, Hong Kong, and Macao. Flexible decision making, short communication lines, greater involvement of citizens and companies in governance structures, and geopolitical importance for major states are among the most mentioned advantages (Streeten 1993).

Before turning to Curaçao and the Netherlands Antilles, we will need to take a look at the environmental dimensions of small island developing states. According to Hein (1990a) and Briguglio (1995), most small islands are characterized by their limited natural resources and their fragile environment, and face environmental problems that are related to their geographic, natural and development features. The Caribbean island states all have a fragile ecological balance between their terrestrial and marine ecosystems, which has been put under increasing pressure by the growth in economic activities, especially in connection with tourism, oil-related activities and construction (McElroy et al. 1990). Small island states also share a common position regarding climate change: they barely contribute to it but are among the first to suffer from sea level rise, increasing storms or a slump in sunbathing tourism, mainly because they lack the resources to protect themselves against these consequences (UNEP 1994). This led McElroy and his colleagues (1990) to summarize the challenge facing micro-states in the years to come as "to create, within the precarious context of current micro-state political economy, supportive incentives, decision-making institutions and management practices that will restore depleted resources and stabilizes sustainable uses."

Curaçao and the Netherlands Antilles in a Global Context
Curaçao can be seen as a prototype of a small island development state, with a size of 444 square kilometers, some 150,000 inhabitants and a GDP of US $1.3 billion. The same is true if we consider the Netherlands Antilles, five islands of which Curaçao is the largest, as a whole: together, they have a size of 800 square kilometers, some 197,000 inhabitants, and a GDP of US $1.8 billion. Although neither Curaçao nor the Netherlands Antilles, being part of the Kingdom of the Netherlands, are nation-states in the formal legal sense of the concept, they do have all the characteristics that are so often attributed to small island development states. The formal "nation-state" character is taken flexibly in most of the literature on this subject. In the list of 79 territories which Bray and Parker (1993) use to "define" micro-states, the

Falkland Islands, Gibraltar, Greenland, and the Netherlands Antilles appear along with Surinam, Belize, and Gabon. I will use a similar flexibility.

The open economy of Curaçao, its concentration on only a few economic sectors and the dominant position of some transnational corporations are among the essential elements that characterize the environmental deterioration and limited pollution control on this island. The economy of Curaçao is largely dependent on the oil and gas sector, tourism and financial offshore services. The former two sectors are particularly relevant for an analysis of the breakdown and preservation of the environment in a globalizing economy.

Small Islands and a Global Oil Economy

For more than 80 years Curaçao has been the host of a powerful oil industry. From 1915, when the transnational company Shell decided to build a refinery on this small island, up to the present day, now this so-called Isla refinery is owned by Curaçao but leased and exploited by the Venezuelan state-owned oil company Petroléos de Venezuela S.A. (PDVSA), the local economy has to a major extent been dominated by the oil sector. The importance of this oil sector for the local economy has made the island greatly dependent on first Shell, and now PDVSA. The environmental impacts of this old, medium-size refinery and the oil-related activities on such a small island are imposing: severe surface water and ground water pollution, air pollution that especially victimizes the poorer inhabitants of Curaçao living downwind, major soil pollution due to leakage and the dumping of waste over many years, energy inefficiencies, etc. During the last 25 years since these environmental ills of the refinery have become publicly known, the local environmental authorities have not managed to make any major step forward in cleaning up the environmental inheritances of the past, nor to initiate a restructuring of the refinery into more sustainable production patterns.

In a detailed analysis of the negotiation processes between the government of the island and the two major transnational oil companies (Shell and PDVSA) throughout these 80 years, van Vliet (1997) shows how Curaçao was consistently overruled in its attempts to safeguard its economic, but even more so its social and environmental interests. Curaçao was not able to press for major environmental investments and improvements, either in the negotiations surrounding the departure of Shell in

1984,[11] the negotiations on the new contract with PDVSA or the renegotiations with PDVSA between 1991 and 1994. The various rounds only resulted in short-term curative environmental measures that were in line with the immediate economic interests and international strategies of the oil transnationals. The main direct causes of these outcomes are related to inequalities in knowledge and economic power, poor negotiation strategies of the government of Curaçao, the closed-door nature of "oil business negotiations" that shut out pressure from non-governmental interest groups, and the ambivalent role of the Netherlands and Venezuela. The home countries of the two oil multinationals supported the companies on crucial moments instead of assisting a small island that is of some relevance and geo-political importance for both countries.

Behind these direct causes lies the globalization in the oil and gas sector. Decisions regarding the Isla refinery, whether related to economic investments, product differentiation, environmental upgrading, cleanup activities, expansion, or anything else, can only be understood by taking the global oil sector and the strategic company policies into account. The government of Curaçao had little if any influence on these decisions, as it is completely dependent on the economic revenues of the refinery and related activities in terms of employment, contribution to GDP, cheap energy and fresh water.[12] Besides, Curaçao is still formally the owner of the refinery. Some of the other islands of the Netherlands Antilles, with economies dominated by major oil and gas activities, are facing the same kinds of problems. The environmental authorities of Curaçao and those of the Netherlands Antilles as a whole have little power in negotiating and implementing environmental regulations with respect to the refinery. Although they have received some support from especially Dutch environmental authorities, they lack the knowledge, manpower, political influence and backing, efficient and effective internal organization. The "rules of the game," formal or informal, generally work to their disadvantage.

Economic Diversification: Tourism
The tourism sector is less dominant in the economy of Curaçao than the oil sector, and compared to some other Caribbean and Netherlands Antilles islands it is rather modest. It contributes "only" 10 percent to the GDP and some 10 percent of employment is directly or indirectly connected to tourism. Tourism brings in more than 60 percent of the GDP of Aruba and

St. Maarten, two Caribbean islands in the Netherlands Antilles. Curaçao draws tourists mainly from the Netherlands (33 percent), the United States (15 percent), Latin America (22 percent, especially Venezuela), and the Caribbean region (18 percent). While the annual number of tourists that spent at least one night on Curaçao totaled about 100,000 in 1975, the number has risen to over 200,000 in 1995, and there are plans to raise it up to 500,000 by the year 2012. The environmental effects[1] of the growing importance of tourism, and of attempts of the economic authorities of Curaçao to stimulate tourism, do not sound unfamiliar: an increase in construction activities in sensitive seashore areas, intensified occupation and privatization of scarce public space, disturbance of fragile coral ecosystems, heavier burdens on the ground water system and a disruption of territorial ecosystems (Abeelen 1997). Seeking to analyze how globalization tendencies intersect with these environmental hazards and attempts to control them, we need to fix our attention on the internationalization in both the tourist industry and tourism competition, the parallel and diverging interests of tourism and the environment and the developments in private-public partnerships in policy making in the tourism sector.

Tourism is widely (and also in government circles) seen as an attractive alternative for Curaçao in its attempts to develop and further diversify the economy in order to make it less dependent on the oil and gas sector. As a small Caribbean island it has the "tropical flavor," although it is less well endowed with the typical features of beautiful white beaches and exuberant nature. Two strategies are pursued in order to compete successfully with other tourism resorts in the Caribbean: redesigning the island to match the standard pictures in the glossy magazines and at the same time driving home the point that Curaçao is more than just a tropical island with palms and beaches: it has a local culture, typical architecture and original natural ecosystems (rather dry with very few trees). The first strategy means, among other things, that beaches are created where they did not originally exist, that huge amounts of scarce fresh water are spent on growing palm trees and maintaining golf sites, and that major investments are made in international marketing to promote this tropical outlook. The second strategy emphasizes the local originalities, the quietness, the quality of services and the friendliness of the local people. In view of the fierce international competition in the region, foreign investors need to be attracted in order to safeguard the level of quality. Major international hotel chains (Van der Valk,

Holiday Inn, Hilton, Ramada, Hyatt) are actively invited to invest in Curaçao in order to attract Western tourists, often at the expense of local investors and hotel owners. Foreign investors encounter few environmental constraints as they are seen as vital allies in international competition, and spatial planning and protection of natural areas are brushed aside in cases where that is necessary to please them. International investors often negotiate tax holidays and price reductions for water and energy. The same is true for the international cruises that need to be coaxed to make a stop in Curaçao's capital Willemstad and not in one of the islands next door.

Still, there is one big difference compared to the oil and gas sector: the tourist sector does acknowledge that it is in its own long-term interest to look after environmental quality, although its definition of environmental quality does not always tally with that of, say, local NGOs or local environmental authorities. The protection of coral reefs, the reduction of industrial pollution,[14] clearing away scattered litter and the protection of terrestrial nature reserves are among the priorities actively pursued by the Curaçao Tourism Development Board (CTDB), a semi-independent private organization that has taken over most of tourism policy from the local government since 1989. The privatization of tourism policy via the CTDB is primarily motivated by the poor performance of the government in developing tourism in Curaçao in the 1980s. The development of tourism, its international competitiveness as well as the attention devoted to some of the environmental aspects of the tourist industry appear to have benefited from this "move to the market."[15] This CTDB has developed a master plan for the next 20 years (CTDB 1993) that is pivotal in Curaçao's tourism policy and will have major effects on other policy areas (regarding the refinery, spatial planning, etc.) as well, provided it is carried out. The position of the CTDB as an intermediary between private (often foreign) owners of tourist businesses and public government has given it the possibility to emphasize the importance of a sustainable tourism sector, albeit defined in its own terms. But the execution of this policy is severely delayed and obstructed owing to its dependence on both private investors and fragmented public agencies. Cooperation between the CTDB and various administrative agencies, each with their own interests and logics of operation, has proved rather difficult. In addition, regional coordination of tourism development in the Caribbean area shows no signs of coming off the ground, partly because competition is too fierce.

The Limits of Small State Governance

The conclusions to be drawn from the developments of these two sectors in a globalizing economy seem rather straightforward. Small island development states are in point of fact limited in their possibilities to direct their economic development into more sustainable directions. Both in the oil and gas sector and in tourism, global economic developments, transnational companies and major states appear to be all-powerful, reducing the degrees of freedom for small states to almost zero. But this is only one side of the story, however great its impact on Curaçao. The oil terminals located on another island of the Netherlands Antilles, St. Maarten, owned and operated by a major American company, show another side: a relatively modern proactive industry with an adequate environmental license and good cooperation with the local environmental authorities.

Shortly after the signing of the new lease contract with PDVSA in 1984 already, voices could be heard saying that Curaçao should become less dependent on the refinery, and stimulate investments in new (industrial and other) sectors rather than continuing to support the refinery that had such a major influence on the economy and ecology. Even the prime minister of the Netherlands Antilles stated in 1994 that "maybe in 10 years' time it will prove possible to set up an alternative industrial structure" (quoted in van Vliet 1997, p. 52). The large amount of capital that Curaçao has put into the refinery could very well be invested more profitably (in terms of employment and contribution to GDP) in other economic sectors that have less devastating effects on the environment. The same discussion on investment priorities can be identified regarding the tourism industry. Suggestions in this debate are less drastic and attract less attention, owing to the lower profile of the tourism industry, its relatively limited economic importance, its more fragmented character and its smaller environmental impact. The state is often looked upon as the crucial institution to redirect the economy toward more sustainable sectors in the future. Two factors are repeatedly raised in discussions on more active state governance in bringing about future sustainable developments: the internal organization of the government and international interdependence and cooperation.

Numerous consultancies and governmental commissions have already tried—mostly in vain—to reorganize and improve the performance of the administration of Curaçao and the Netherlands Antilles, often from a budget-cutback perspective. These unsuccessful attempts demonstrate how inert some of the administrations in micro-states are and how closely they are entwined with the other economic, political and cultural institutions that are so characteristic of these states. This picture does not leave much room for optimism.

A number of aspects seem both essential and feasible (Mol and van Vliet 1997b) in any attempt to develop a program of political modernization on Curaçao to overcome the institutionalized unsustainability of state governance: a functional separation of policy-making and policy-implementing agencies; the separation of conflicting interests embodied in one individual leading politician; transparency and democratic control of (para)statal agencies and companies; qualitative improvement of the administrative authorities; further centralization of environmental policy tasks and responsibilities at the level of the Netherlands Antilles; and a further integration of environmental policy in other sectoral policies. Having learned from the numerous attempts of international consultancies trying to reorganize the governance structure, we also have to conclude that models of (environmental) governance and reform for Curaçao cannot be mere copies of those prevalent in the triad countries. Local historical, cultural and institutional characteristics of such a small island development state need to be taken into account.

The second factor in strengthening the government's role in environmental reform hinges on international cooperation. International cooperation—both between state agencies and between non-state organizations—is mentioned time and again as being essential in strategies to overcome the unsustainable development path of micro-states that are so tightly bound up in globalization. Until now, most of Curaçao's international cooperation in the field of the environment has been with private and public agencies from countries outside the Caribbean region, most notably from the Netherlands, the United States and—to a lesser extent—Venezuela. Stronger regional (environmental) cooperation, for instance within the CARICOM, might prove to be more rewarding in the long term. The problems experienced are more similar and sometimes even interdependent, solutions will be less focused on ill-suited high-tech systems, economies of scale can be accomplished, and political cooperation can prevent these states from being overlooked, overruled or marginalized in international frameworks and MEAs. Such regional harmonization and collaboration has come about within the framework of the Alliance of Small Island States AOSIS during the UNCED 1992, the Barbados 1994 conference and the negotiations within the Framework Convention on Climate Change (Paterson 1996; Grubb et al. 1999), and with some success. It is exactly political globalization that seems to be part of the answer to the failures of environmental reform in the micro-states suffering from strong economic dependence and powerless national governments.

Predatory States in Africa: Kenya

The African continent, and especially the sub-Saharan region with the exception of South Africa, has long been considered the most difficult continent in terms of economic development. Though several countries in Asia and Latin America have shown accelerated economic and industrial development throughout the 1970s and the 1980s, Africa seemed to be moving in the opposite direction. In Africa "industrial production has stagnated or declined in many countries over the past decade" (Lall 1992). Representatives of major international development organizations have repeatedly expressed their pessimism as to the economic future of Africa, although their evaluations where often rather superficial and one-dimensional. Kenya has for some time been considered one of the few exceptions in Africa. Its economic development, political stability and integration into the Western global world were often cited in order to point out the possibilities for the African continent. The last decade or so has shown that Kenya is more typically African than most commentators had believed or hoped in the 1970s and the 1980s.

Industrial Development and Environmental Deterioration

Kenya is among the most industrialized sub-Saharan African countries.[16] As most of the African states, the Kenyans saw industrialization as vital to the further economic development of the country on becoming independent. The contribution of industry to its economic development (in terms of GDP) has increased since independence in 1963 to some 17 percent by the mid 1990s, though it reached 21 percent in 1980. The average annual growth rate of industry also decreased in later years, from 3.9 percent over the years 1980–1990 to 1.5 percent over the years 1990–1995, but remain above the average of sub-Saharan Africa (World Development Report 1997). According to the same source, the (formal) industrial sector in Kenya employs some 7 percent of the registered labor force, as against agriculture and forestry which employ some 80 percent, about the average of sub-Saharan Africa. Out of over 700 medium- and large-scale enterprises operating in Kenya, over 200 are part of foreign multinationals (Percival 1996). Most of the larger industries are located in the two main cities, the capital Nairobi and the port town Mombassa, despite recent efforts of the Kenyan government to diversify industrial siting toward rural areas. Besides these larger enterprises, Kenya, like most other African states, has a large number of small-scale—often informal—enterprises that are believed to employ some 30 percent of the total labor force (compared to over 45 percent in countries such as Nigeria and Sierra Leone). Informal industrial enterprises are said to employ some 6 percent of the total labor force. Enterprises with less than 10 employees, usually classified as informal *Jua Kali*, are well represented, while the "missing middle" is often believed to be an indicator of the weakness of indigenous industrial entrepreneurship (Lall 1992, p. 108). Still, Kenya is considered to be one of the more promising economies in this respect, with a group of Asian entrepreneurs being especially successful. Poor industrial export performance is common among most African economies and Kenya is no exception, with less than 20 percent of its exports consisting of manufactured goods (World Development report 1997).

Throughout the years, governmental industrial policy in Kenya has been similar to that in most African countries: a combination, or rather alteration, of two distinct strategies. The official government strategy of African countries has for a considerable time been focused on external conditions: attracting more aid, raising export prices and reducing the debt problem. International institutions such as the World Bank and the IMF, on the other hand, have focused—especially since the early 1980s—on stabilization and structural adjustment policies (Lall 1992; Clapham 1996). The partially successful import substitution policy pursued by Kenya in the 1970s was changed to a structural adjustment policy in the National Development Plans of the 1980s, mainly as a result of pressure from the World Bank and other external sources to reduce protectionism of Kenyan industries.[17] The limited economic success forced Kenya into the international arena, leading to indebtedness and a weakening of the state's control over markets and prices. It was from then on that the informal sector, which has the subversion or evasion of state control as its common feature, became the major growth area. Since then Kenyan policy has been mixing both ingredients by encouraging direct foreign investments (via tax policies, licensing, granting monopolies, freedom of locational choice, etc.), stimulating domestic industries ("Kenyanization") by financial and infrastructural efforts and continuing (be it in vain) to control and plan industrialization and economic development (Ombura 1996). The main policy goal of the Kenyan government in the 1990s was set out in Sessional Paper 2 (Kenya Government 1996): "national policies and strategies that will lay the foundation for the structural transformation required to enable Kenya to join the league of Newly Industrialized Countries," and this goal was further elaborated in the eighth Kenya National Development Plan (1997–2001). The government perceives its role as creating an enabling environment, both for large and foreign industries and for small-scale *Jua Kali*, to develop within strict (environmental) boundaries. But while Brautigam (1994) and Evans (1995) have demonstrated the importance of a strong state in the maturation of small-scale industrial activities toward an industrial society, the Kenyan state—along with other African states (Lubeck and Watts 1994)—seems unable to achieve that goal.

Industrial development in Kenya is accompanied by serious environmental pollution, most notably in the major urban and industrial centers of Nairobi and Mombassa. Hardly any industrial solid waste, up to some 550 tons per day in the city of Nairobi, is collected separately from domestic solid waste (Situma 1992). Almost all industrial wastewater flows into the Nairobi River, either directly or via the sewer system, adding to the health hazards caused by untreated domestic wastewater. Air pollution is often above the international WHO guidelines. As industry is generally dispersed throughout the residential areas, environmental and health risks are serious and often difficult to control or prevent. A clear difference should of course be noted between the 700 major industrial production sites and the numerous small manufacturing enterprises. Whereas the former have some access to financial resources, modern technology and management methods, environmental knowhow, and political influence, the small businesses are generally deprived of such resources, making environmental reforms extremely difficult (Frijns et al. 1997). Kenya's contribution to global environmental problems is limited: its greenhouse gas emission are negligible (5 million metric tons, 0.18 ton per capita; wood fuels contribute 75 percent to these emissions, while industry contributes 8 percent (Gupta 1997)), as are chlorofluorocarbon emissions.

Governmental Environmental Policy

The emergence of national environmental policy institutions and activities dates back to the late 1970s, when the National Environmental Secretariat was established within the Ministry of Environment and Natural Resources. Industrial pollution control was given some priority at the time, and environmental impact assessments were considered to be the most appropriate tools for combating industrial discharges and evaluating new industrial facilities. In reality, EIA stagnated, owing to poor implementation and the weak inter-departmental position of the National Environmental Secretariat. Although the National Environmental Secretariat tried to strengthen its position with respect to the other ministries, using the Inter-Ministerial Committee on the Environment as a forum to strategically promote its own goals, it never succeeded in becoming a powerful player among the other ministries. Only after the Brundtland report and the 1992 UNCED conference was industrial pollution control successfully put on the national agenda again, followed by the adoption of the National Environmental Action Plan in 1993. The relationship between the old National Environmental Secretariat and the new group set up to carry out the National Environmental Action Plan, both working within the same ministry, remained unclear until—with foreign assistance—the National Environmental Management Bill was passed in the late 1990s. This new law provides a comprehensive institutional framework at the national, district and local levels. The National Environmental Council (representing the main governmental and private interests) and its executive branch, the National Environmental Management Authority, are supposed to function as the supreme bodies for policy formulation and implementation.

Although on paper environmental regulation and policy-making instruments seem rather solid (perhaps except for air pollution[18]), there are several causes working against an influential environmental reform strategy in practice. Understaffing, lack of personal and institutional capacities, limited monitoring potential, poor environmental facilities and organizational structures, competing competencies of governmental agencies, and failing enforcement are among the shortcomings, which do not sound too unfamiliar after the previous two case studies. Nevertheless, some differences need to be underlined. Whereas in Vietnam the institutional structure to deal with the environment is clearly in the making, and Curaçao encounters some of the specific difficulties that small states face when confronted with major international polluters, Kenya has been active in the field of environmental regulation and control for a considerable period of time. It has signed various international treaties on global environmental problems, such as the Framework Convention on Climate Change.[19] Moreover, it has profited from major international support in developing its environmental institutional structure, not least because the UNEP has its headquarters in Nairobi. The explanation for its rather poor record in environmental management can partly be found in domestic developments (as touched upon above), but partly—and for our purpose more interestingly—also in the relationship between processes of globalization and the Kenyan nation-state.

Globalization, Industrialization, and Environmental Control

In his study on Africa and the international system Clapham (1996) convincingly argues that Africa has hardly been directly affected by the dynamics of globalization. The globalization of economic capital, for one, passed Africa and Kenya to a major extent, "because there were few places where transnational corporations could find safe and potentially profitable investment opportunities" (ibid., p. 25). Castells (1996, p. 135) joins him in noting that the "new global economy does not have much of a role for the majority of the African population in the newest international division of labor. Most primary commodities are useless or low priced, markets are too narrow, investments too risky, labor not skilled enough, communication and telecommunication infrastructure clearly inadequate, politics too impredictable, and government bureaucracies inefficiently corrupt. . . ." As a result, the African states have become increasingly structurally irrelevant for the global economy. Analyzing investments of UK transnational industrial companies in 14 sub-Saharan states including Kenya, Bennell (1995) confirms these observations and shows that foreign direct *disinvestments* are the general picture since the mid 1980s, despite concerted efforts by African governments and international organizations to push for structural adjustment. If anything, economic globalization leads to a further economic marginalization of Kenya with respect to the world economy. In the same line, the flows of information that are so characteristic of the era of globalization (whether related to financial markets, the Internet, or environmental data and information) have had a relatively minor impact sub-Saharan Africa.[20] Thus, although it looks as though globalization does not touch

Africa directly, it does alter Africa's external environment and thus produces drastic indirect effects. For one thing, it is undermining the position of the Kenyan state, as Clapham (1996) argues. The combination of ongoing economic globalization and poor economic achievement have made Kenya and other African economies and states more dependent on global markets, foreign loans and international aid,[21] and the global actors "in charge" of them. The stagnating growth or even decline in primary and industrial production (and the decreasing terms of trade) thus undermine the state's autonomy with respect to the external, global world. But at the same time it has affected its authority with respect to the domestic players. In order to secure their patronage system (and thus their power) the state, or more precisely the ruling elites (Castells 1997b, pp. 82–105), extracted more and more resources from the local economy and reallocated them to a small group on which they depended. This reduced the legitimacy of their rule considerably among the majority of the domestic interest groups. Hence, globalization constantly threatens to undermine the sovereignty of the African state, and maintaining their sovereignty especially over the domestic interests was and is a constant struggle for African rulers, resulting in the state concentrating on survival rather than on navigating its society and economy, for instance into more environmentally sound directions.[22] The withering away of the state and its diminishing sovereignty in various fields has not only led to the emergence of weak predatory states and booming informal sectors. It has also created room for non-state actors to move into some of the traditional state tasks and strengthen the influence of international actors and dynamics. What does this mean for environmental pollution and control?

The weakness of the Kenyan state and its constant need for financial resources to survive, also affects the functioning of environmental management and control and to some extent explains the poor results in this area. Though the majority of the smaller and informal industrial firms are only incidentally visited by environmental inspectors, the large and especially the international industrial enterprises regular see environmental authorities. But large foreign firms do manage to avoid serious state intervention because they have the resources and expertise to fulfill the environmental demands and they have access to political or financial resources. However, this does not mean that they always perform according to sustainability standards. Environmental priorities are often low and their performance is generally below the standards of countries in the triad. The smaller, informal firms, meanwhile, have virtually no financial, political or technological resources to deal with environmental inspections, so they become easy targets for badly paid environmental officials in search of some additional income. Smaller enterprises are also overlooked in all kind of governmental support programs, although during the last decade governmental and especially non-governmental organizations have begun to devote more attention to the informal sector as a source of income, employment and economic growth, and some environmental support programs have been set up (Frijns and Malombe 1997). And even in the unfavorable institutional environment of a predatory state, small-scale results in environmental reform can be achieved on the basis of local cooperation and trust between public and private organizations, as various of our investigations have shown (Wanyonyi 1996; Onyango 1997; Wasonga 1999). But on the whole and in the environmental field, the Kenyan state shows its "predatory" face: limited collaboration with and support for such industries and industrial organizations and rather seeing them as sources of income. One of the consequences of the rather poor environmental performance of the Kenyan state is that environmental reforms are increasingly effected by international organizations and international cooperation.

Increasingly, international donor and national bilateral assistance programs not only set the agenda and priorities, but also design and implement environmental reform programs. Formally, these donor programs are carried out in close collaboration with governmental organizations, and institutional capacity building is one of the standard goals. Nevertheless, the degree to which foreign agencies "take over" national environmental policy making and implementation is astonishing. Every major environmental policy innovation is accompanied for a considerable period of time by major foreign dominance and the programs that are put in place more than occasionally collapse within a few years after collaboration ends. These mixed experiences of foreign assistance targeted at governmental organizations have—more recently—resulted in a tendency toward de-stating.

International assistance for environmental reforms is increasingly focused on non-state actors and agencies, as is the case for assistance in other fields. As long as international assistance was provided via the state and associated with it, it was perceived as state-strengthening. It provided the state with the resources to bind national interest groups to it and thus played a part in the state's survival strategy. But the increasing movement of international assistance or ODA toward NGOs (or directly to municipal authorities), instead of taking the national state as the official and/or actual counterpart, further weakens the state, especially in those cases where NGOs are already trying to move beyond the state using local resources. The "environmental de-stating" in transnational North-South relations, with international and non-governmental organizations fulfill the responsibilities of the North and NGOs and local communities taking over the Southern part, has a long history in Kenyan nature conservation. But state-to-state relations are also becoming less vital in initiatives to address more urban and industrial environmental problems in Kenya and other African states. In Kenya the Undugu Society (an NGO based in Nairobi) takes over several traditional state tasks in the field of small enterprise development and urban environmental protection in Nairobi and is strongly supported by international funds. So is the Green Towns project, where international donors closely cooperate with local communities in greening towns, almost completely beyond interference of the national state. These "subpolitical" developments may be fruitful where environmental and economic interests can be put in a win-win situation with the help of major financial resources from the outside, but success is difficult to imagine in situations where the two are in conflict and "tough" environmental measures need to be taken. It would appear that the national state still has an essential role to play in such circumstances, and also with respect to implementing multilateral environmental agreements.

Comparative Conclusions

If we are to draw conclusions from these case study countries, it will rather be on the diversity than on the similarities: the diversity of the ways in which globalization processes affect the local dynamics of environmental degeneration and environmental reform, the diversity in the degrees of "environmental sovereignty," and the variety of institutional frameworks and mechanisms by means of which the national political entities seek to safeguard environmental quality.

Vietnam is only just opening up to the global economy, and the state is still very much in control of the national economy. This implies that "conventional" state-controlled environmental reforms are—and will be for some time—of major importance in improving the environmental performance of industrial processes and products, whether that of local industries or transnational companies investing in Vietnam. The other two case study countries seem to be in full contrast with the case of Vietnam, albeit each in different ways. As small island development states, Curaçao and the Netherlands Antilles are open economies completely dependent on global economic dynamics and on the economics and politics of a few more powerful states: the United States, Venezuela, and the Netherlands. The small state has little control over the local economy, which is reflected in its inability to redirect its major economic sectors into more environmentally sound paths. Important environmental reforms will primarily originate in the market, where local consumers, customers and producers push for environment-induced changes of production practices, or international economic players and markets induce environmental improvements. In Kenya, the state has equally limited—actual or potential—influence on environmental reform, but that is not so much because of the full integration of its economy in the world order. Consequently, widespread environmental reforms are not as likely to be triggered or carried by national or international market actors and dynamics. NGOs and decentralized state organizations—often supported to some extent by international agencies—are the ones to push for an environmental orientation in an as yet stagnant industrialization process.

Evidently, therefore, the distinct institutional traits of a country need to be taken into account in any analysis of the potential and actual environmental threats and opportunities of globalization processes. In Vietnam, globalization can bring major technological improvements to its rather old industrial production systems and thus produce major environmental gains, while it can also help to create a stronger—internationally backed—civil society pressing for environmental reforms. The threats, on the other hand, lie in a weakening of the state following deregulation and liberalization demands of international capital, while a regional supra-national state-like structure or an active civil society are absent. As for Curaçao, political cooperation among similar small island development states, especially in the region, can fortify its position, not only in international environmental negotiations and arrangements, but also with regard to regional capital looking for cheap production conditions. In addition, Curaçao has everything to gain in environmental terms from a greening of global capital and markets. Its heavy dependency on the international economy or market and a few major nation-states is at the same time its weakness, since a major part of the trends in the global economy and politics is still not going green. Compared to the other two, Kenya has only little to gain in environmental terms from major global economic developments turning greener. Its integration in the world economy is limited, and it will sooner be the global connections in the political and civil society dimensions that can push Kenya into more environmentally sound directions. A reconstruction of the state might be a viable option. It would require both a bottom-up movement, with local authorities acquiring new capacities and building new institutions with the help of international assistance, and a top-down movement via multilateral environmental agreements and regimes and the attendant side programs for developing states.

9

Ecological Modernization and Global Environmental Reform

I began by outlining what seems to be a paradox in the contemporary discourse on globalization and the environment. On the one hand, the environmental debate on the relation between the institutions of modernity and the sustenance base is clearly moving toward a new consensus. Whereas in the 1970s and the early 1980s the damage done to environmental quality by the institutions of modernity held center stage in environmental discourse, a more balanced position has moved to the forefront since the mid 1980s. Technological developments, economic markets and mechanisms, nation-states, and political institutions and arrangements were increasingly recognized, not only for their detrimental effects on the environment, but more and more for their contributions toward protecting the environment. Institutional developments and transformations in Western societies are more and more informed by environmental interests and considerations, even if economic dynamics and interests are still far more powerful. The Ecological Modernization Theory in particular has tried to conceptualize and understand these environment-informed institutional transformations and dynamics. The growing interest aroused by this theoretical framework is evidence of a general understanding that, in this respect, the late 1980s and the 1990s are indeed different from the 1970s and the early 1980s.

On the other hand, I concluded my analysis of the various globalization theories that have emerged since the early 1990s by saying that the environment has rarely been more than a minor issue in globalization debates up until now. And in those rare instances when the environment did appear on the stage of globalization studies, the common view put forward by most scholars was a rather negative one: globalization processes and trends add to environmental deterioration, to diminishing control of environmental problems by modern institutions, and to the unequal distribution of environmental consequences and risks between different groups and societies. The dominance of economic (that is, capitalist) globalization processes is often believed to be the root cause of these detrimental environmental effects. Global political institutions, arrangements, and organizations and a global civil society are believed to be lagging behind.

Two explanations for this environmental "disarray" have been put forward in the previous chapters. I will add a third and related explanation in this chapter.

The first explanation is that environmental scholars have overlooked or underestimated the impact of globalization dynamics on environmental reform. The argument made in this book is that the positive signs of modern institutions taking up environmental considerations and "interests" have for a considerable time been associated primarily with the "old" institutional order of high modernity. The "new" institutional order that is slowly evolving under conditions of globalization was to some extent neglected or undertheorized. Ecological modernization studies and related analytical frameworks focused too much on the national economies (mainly those of the triad countries) in their analyses of the institutionalization of the environment in market dynamics and mechanisms, and showed too little awareness of the growing economic complexities and global interdependencies. In their assessments of the major political modernization trends in some of the Western nation-states faced with environmental challenges, eco-modernizationists initially overlooked or underestimated the fact that in an era of globalization the nation-state is losing part of its central position. The eco-modernizationists emphasized the incorporation of environmental considerations in technological designs of products, production processes and socio-technological systems. Meanwhile, however, they tended to lose sight of the upscaling trends and increasing global complexities and interdependencies of these technological systems and networks, but also of the feelings of anxiety, fear and lack of control experienced by citizens and consumers in conjunction with these "globalizing" technological complexes.

The second explanation of this disarray relates to ill-informed and poorly equipped globalization scholars (both hyperglobalization adherents and the more critical transformationalists; see Held et al. 1999, pp. 3–10). Like most of the sociologists that aimed to analyze the modern societies of the

1970s, contemporary globalization scholars are not sufficiently equipped or trained to be able to move beyond the truisms of modernization and globalization processes causing environmental deterioration and find their way toward more refined analyses of the kind of transformations and reforms taking place in modern society's relationship with nature. Reacting against neo-liberal economists who appear to almost unconditionally celebrate globalization processes for their economic but also environmental benefits (Ohmae 1995; OECD 1997), sociologists, political economists, and anti-globalization activists all too easily reduce globalization to unfettered global capitalism. They see a globalization virtually unhindered by correction mechanisms—economic, political, or cultural—to counterbalance its various disastrous side effects. Thus, while the environment is still anything but the center of the social scientists' attention, the negative environmental side effects of globalization are either trivialized by neo-liberal economists or taken for granted by their opponents. Only global environmental movements are seen by the latter group as marginally and peripherally counteracting the dominant, almost apocalyptic trend of global environmental decay. The number of more refined analyses of processes and mechanisms linking globalization to environmental deterioration as well as reform is indeed very limited.

A third explanation for this disarray relates to contrasting evaluation perspectives, as put forward in chapters 4 and 5. This study tends to emphasize the impact of these global developments and mechanisms on restricting global capitalism's devastating environmental consequences. Other scholars contradict this view, as I noted in chapter 4. The discussion between neo-Marxist-oriented Treadmill of Production perspectives and Ecological Modernization ideas (Pellow et al. 2000; Schnaiberg et al. 2001; Mol and Spaargaren 2000, 2001) sheds more light on the contradictory evaluations of the continuity of global capitalism and the discontinuities in emerging environmental transformations. Neo-Marxists have condemned the Ecological Modernization perspectives for their limited focus and their failure to get at the "roots of the environmental crisis" (Pepper 1984). Four major points can be raised to clarify the differences between the two schools of thought in evaluating the attempts at global environmental reform (table 9.1):

• Ecological Modernization studies concentrate on "environmental radicality" rather than on "social radicality." That is, in their assessments of existing patterns of change-in-the-making Ecological Modernization

Table 9.1
Environmental change versus economic continuity.

	Treadmill of Production	Ecological Modernization
Kind of radicality	Economic radicality	Environmental radicality
Environmental improvements	Absolute sustainability	Relative improvements
Assessment of environmental change	Window dressing	Real changes
Relation between changes analyzed and changes proposed	Weak relation	Strong relation
Main emphasis	Institutional continuity	Institutional transformations

perspectives tend to focus on the contributions to environmental reform, and not primarily on the effects of these changes in terms of various other criteria. "Small" deviations from the existing institutions and practices can produce substantial environmental improvements, just as "big" changes in terms of a radical or fundamental reorganization of the economic relations of production can have limited environmental benefits (as we know from formerly communist Eastern Europe). Ecological Modernization is first and foremost an environmental social theory, analyzing the environmental origins and environmental consequences of social change. It does not deny that social change can be evaluated in terms of all manner of social criteria (distributional effects, gender issues, etc.), but these are separate, relatively independent criteria and should not be put into the same category with environmental criteria (as is advocated in some versions of the notion of sustainable development). In principle, ecological modernization theorists might very well come to the conclusion that green capitalism (in whatever form) is possible from an environmental point of view, but that the social consequences would be so dramatic that a move in that direction would be very unlikely and undesirable. In that sense, ecological modernization is a more focused—others would say more restricted—theoretical framework than the Treadmill of Production perspective. Treadmill of Production scholars seem to be primarily interested in changes that involve a transformation of the capitalist or treadmill character of production and consumption. The hypothesized one-to-one relationship between the "relations of production" and environmental disruption causes them to count changes as significant only if they undermine the treadmill, which is consistent with the Treadmill of Production scholars' all-embracing notion of capitalism. The notion contained in the Ecological Modernization Theory of a growing autonomy of the ecological sphere contradicts such an automatic coupling of the economic sphere (or "relations of production") and the environmental sphere.

• The analysis of change in the two perspectives differs in terms of what might be called "absolute" (Treadmill of Production) and "relative" (Ecological Modernization). Criticizing the Ecological Modernization Theory for its rather naive ideas on environmental improvements, Treadmill of Production scholars claim that all—or the overwhelming majority—of production and consumption practices are still governed by treadmill logics, and that ecological, environmental, or sustainability criteria will seldom if ever be dominant in the organization and design of production and consumption. In my view, however, this is not so much what contemporary ecological modernization theorists will or should deny, and I certainly have not done so in this book. I would agree that treadmill (or economic) criteria and interests play a crucial and dominant role in organizing and designing global production and consumption, and that they will probably always remain at least as important as ecological or other criteria (Mol 1995, pp. 28–34). But the innovation is that ecological interests and criteria are slowly but steadily catching up with economic criteria. And this is also true on a global scale, if not to the same extent and in the same way all over the planet. Compared to some decades ago, environmental interests can no longer be ignored and increasingly make a difference in organizing and designing production and consumption. This innovation does not mean that economic or treadmill logics have or will become subordinate, or that economic production and consumption have or will become completely sustainable (whatever that may mean). In that sense, the Ecological Modernization Theory looks at relative (but significant) changes into more environmentally sound directions, in contrast to the "absolute" sustainability on which Treadmill of Production scholars focus.

• There is a major difference between the two perspectives in their assessments of the environmental changes that have been set into motion from the late 1980s on: window dressing (Treadmill of Production) versus structural changes in institutions and social practices (Ecological Modernization). It goes without saying that empirical evidence to underpin either of the two can easily be found and constructed. After all, the large variation in data sets, criteria, variables, and time intervals rules out the possibility of any "objective" final answer or conclusion. Treadmill of Production scholars insist that they see no real and lasting environmental improvements and therefore define all environmental initiatives and institutional changes as window dressing: nothing new to report. I would, however, still claim that an assessment of environmental transformations in terms of window dressing seems to bypass the differences that exist between the current institutionalization of the environment—regardless of all the shortcomings and limited successes—and that of the 1970s. To quote Castells (1997b, p. 336):

"After all, if nothing is new under the sun, why bother to try to investigate, think, write, and read about it?"

• A distinction should be made between the nature of the changes advocated by the two frameworks. Both the Treadmill of Production perspective and the Ecological Modernization perspective contain analytical as well as normative and even prescriptive dimensions. (See chapter 3 above.) This means that both perspectives analyze contemporary processes of social continuity and change but also seek to contribute to the development of normative political trajectories of transformation that ought to take place in order to turn the tide of environmental destruction. Most Treadmill of Production studies report a major gap between the quite advanced and detailed theoretical analyses of the immanently destructive character of the treadmill of global capitalist production and the suggestions made for concrete trajectories toward social change. Analyzing the roots of environmentalism and the environmental crisis from a neo-Marxist position, David Pepper (1984) fails to put forward a feasible program of social change targeted at these roots. He ends up with environmental education as the main strategy. Equally, James O'Connor's latest book (1998) and Walter Goldfrank's (1999) World-System Theory volume contain detailed and refined neo-Marxist analyses of the destructive pattern of the capitalist world economy, but rather "meager" and utopian countervailing strategies for environmental reform. It seems to me that the strategies for change developed within the Treadmill of Production perspective and related perspectives have not been improved and refined in step with their analyses of environmental disruption, and that they are founded only marginally on existing patterns of social transformation and thus have a highly "utopian" character. In contrast, within the Ecological Modernization Theory there is a closer link between the analyses of existing changes in the making in the main institutions and social practices, and the design of "realist-utopian" (Giddens 1990) trajectories for environmental reform for the near future.

In my attempt to clear up this globalization-environment confusion, I have aimed to counterbalance the generally pessimistic views held by environmental scholars and activists—without being trapped in neo-liberal trivializing and denials—by presenting analyses of environmental reform occurring in conjunction and linked with globalization processes. At the same time, I have aimed to re-interpret and reformulate the Ecological Modernization Theory so that it may deal more adequately with the new institutional order under conditions of globalization. The aim, therefore, is not to deny or falsify the detrimental environmental side effects of globalization processes (as they are real and by no means insignificant). The focus on the actual dynamics of global environmental reform follows neither from some—desperate—hope, nor from the conviction that academic environmental studies must have a constructive attitude toward environmental improvements. Rather, this selective analysis is rooted in the ideas that insufficient light has been shed on the reform side of the globalization coin in environmental discourse and debate, that the most innovative and interesting contemporary social transformations are linked to the design and implementation of global environmental reforms, and not part of a continuing dynamics of global environmental crises, and that contemporary social theory largely fails to clarify these global reform practices and institutional developments, as compared to either neo-liberal economics studies, which deny the existence of structural environmental problems, or social theories on the theme of continuing global environmental deterioration (such as the Treadmill of Production, World-System, and Risk Society schools of thought).

The point that I have sought to make throughout the various chapters is that, although global capitalism has not been beaten and continues to show its devastating environmental effects in all corners of the world, we are moving beyond the era of a global treadmill of production that only further degrades the environment. Powerful, reflexive, countervailing powers are beginning to get a grip on the contradictory developments of environmental reform. The battle of Seattle was an illustrative example of this, because it attacked what is often seen as the most powerful representative of global capitalism: the World Trade Organization. And the attack was such that one cannot expect the WTO to continue its business as usual. What are these emerging mechanisms and actors that try to tame the treadmill of global capitalism in order to save the environment, with major or minor successes?

Taming the Treadmill of Global Capitalism

Political Modernization

Political scientists and international relations theorists, in particular, have concentrated on the construction of global, multilateral or supra-national environmental organizations, institutions and regimes as instruments to contribute to[1] environmental reform of a globalizing world order. Especially since the early 1990s, international relations scholars have seen environmental issues as a new and interesting issue for multilateral actions, institutions and regimes. They have devoted a great deal of attention to the numerous multilateral environmental agreements, most of which focus on one or a limited number of environmental issues (such as Multilateral Agreements on Environment on the protection of the ozone layer, the export of waste, transboundary air pollution, the protection of the oceans, or the Framework Convention on Climate Change). Although they remain a set of "piecemeal arrangements," the expanding number of multilateral environmental agreements are increasingly moving toward common denominators in terms of legal and policy principles (via spillover and other mechanisms), and thus becoming more relevant as building blocks for universal international environmental law and policy. In that sense, they jointly contribute to the emergence of a relatively independent environmental realm in global politics.

Nevertheless, I have argued that, in the end, the regional, originally economic institutions such as the European Union and to a lesser extent the North American Free Trade Agreement are probably of greater relevance for the future taming of transnational capitalism. The "institutionalization of the environment" in these regions has proceeded beyond a level of piecemeal or issue-specific environmental arrangements. The design of overarching political institutions and arrangements, originally intended to further economic integration, increasingly includes environmental protection. The same is true for the economic arrangements, albeit to a lesser extent. In most of the other regions—including the third angle of the triad centered around Japan—similar institutions have until now remained dedicated to trade liberalization and economic integration. And although this environmental inclusion is far from ideal from an environmental interest perspective, most scholars looking for promising developments and prospects in the taming of transnational capitalism are turning their attention to the European Union and to a lesser extent NAFTA. (See, e.g., Group of Lisbon 1995; Held 1995; Martin and Schumann 1997; Beck 1997.) The preference for the European Union above NAFTA as a model for future global governance is due to its relatively strong supra-national institutions (such as the Commission, the European Parliament and the European Court of Justice), which are to a major extent lacking in the Multilateral Environmental Agreements as well as in the greenest trade agreement to date: NAFTA. This makes the European Union unique, not only because it has the supra-national power to counteract the environmental side effects of global capitalism, caused or facilitated by member states and TNCs linked to these states, but because it is also the first experiment in supra-national democratic governance, as advocated so strongly by David Held and his colleagues (Held 1995; Held et al. 1999).

Beyond the Upscaling of National Environmental Arrangements   From an environmental reform perspective, most of the new and primarily political supranational entities are to some extent the equivalent of the national political arrangements on which they are often heavily inspired. Since the late 1970s, various individual nation-states have produced national political arrangements that have had some success in turning structural ecological deterioration into environmental improvements. The basic idea now seems to be that since environmental problems have moved to supra- and transnational levels, in terms of both causes and manifestations, the political institutions and arrangements to deal with them must also be "upgraded" to those levels in order to remain effective: "es handelt sich letzten Endes um die Strategie eines 'Weiter-so' auf gehobenem Niveau" ("In the end, it is a strategy of 'more of the same thing' on an elevated level") (Beck 1997, p. 221).

But there are some serious shortcomings in this rather simple idea of upscaling.

First, in the age of globalization, environmental deterioration has taken on an entirely different aspect as compared to the situation in the 1970s and the 1980s. This change goes much further than a change in scale, and therefore merely upscaling the nation-state institutions and political arrangements for environmental reform to the global level will not do. The dynamics of environmental deterioration and effective reform in an era of globalization are not so much related to geographical scale but to the specific characteristics of the globalization processes that I discussed in chapter 2. The actors involved in triggering political innovations, the legal status, the absence of a sovereign entity and the "democratic" limitations of alternatives for such a sovereign entity, the changing character of capitalism itself, and the disenchantment of science are but a few of the factors which make that supranational or global political institutions must, and in fact do, deviate fundamentally rather than marginally from their national counterparts.

A second and partly related reason is that supra-national, transnational or global political institutions appear to be more successful in combating environmental decay with respect to the developed triad than with respect to the developing countries. There are profound differences between countries in terms of economic development, political and economic integration in the global system, national political institutions and environmental reform capacity (Jänicke 1991), as well as between regions with respect to supra- or transnational political institutions. Moreover, with respect to environmental decision making and implementation, the global political system still depends to a large extent on nation-states. As a result of all these factors, "upscaling" will have different consequences and effects in different parts of the planet. As we have seen in chapter 8, some developing countries are barely "touched" by the emerging global political institutions and agreements aiming at environmental reform, so that they have little to gain from them to alleviate their environmental problems and crises, whether induced by globalization or not. The situation in some parts of the triad looks more promising in this respect, though not enough to simply embrace the (naive) optimistic faith of hyperglobalization adherents.

Third, under conditions of globalization, political arrangements and institutions dealing with environmental reform are no longer restricted to the level of the nation-state system. Decentralized forms of government (such as municipalities or regions) are also appearing on the global stage of environmental politics. Furthermore, global environmental politics, regardless of the level, now also involves actors other than the traditional political agents and institutions. Subpolitical developments—environmental politics involving actors and mechanisms outside the traditional political domains "occupied" by (the system of) nation-states, parliaments and political parties—are interpreted by some as a new answer to environmental deterioration, following some of the typical features of globalization. Nongovernmental environmental organizations have always been at the forefront of environmental reform, but until recently their role in international environmental politics was restricted to, basically, pressuring the traditional political agents—individual nation-states or collectives of nation-states meeting to negotiate environmental agreements—into action. Conversely, the role of transnational enterprises has traditionally been one of either simply causing environmental deterioration or (hesitantly, reactively or even symbolically) complying with reform measures in response to pressures from, primarily, national governments.[2] These traditional patterns seem to be changing: both the agents of "civil society" and the agents of economic interests are beginning to become active and powerful in environmental politics at the sub- and supra-national levels. Such innovations along the lines of ecological modernization can only be understood against the background of a weakening system of sovereign states, the limited achievements of these nation-states, and emerging globalization processes. The following two sections focus on these sub-political innovations, and their role in taming the global treadmill.

Economic Dynamics

One of the major innovations—and at the same time one of the most disputed ones—of the Ecological Modernization Theory and related interpretation frameworks has been the notion that market dynamics and economic actors have a distinct role to play on the stage of environmental reform, and are already doing so in the most developed nations. What is referred to here is not the isolated free markets, the ideal-typical capitalist settings or the short-term-profit-maximizing companies that have no regard for continuity. Environmental reform is coming about in the interplay between economic markets and actors on the one hand, and (organized) citizen-consumers and political institutions seeking to condition them on the other. Such interplay allows environmental considerations, requirements and interests to slowly but increasingly become institutionalized in the economic domain. If such market-induced or economy-induced environmental reforms have come about in national settings, will they also hold in an era marked by globalization, and what would be the difference?

Market-Induced Environmental Reforms   In the previous chapters, I have indeed identified several "economic" mechanisms and dynamics which redirect global capitalist developments and trigger or mediate environmental innovations and reform. As a rule, such economic mechanisms and dynamics do not originate in the economic domain itself.[3] In that sense "market failure" in the provision of common or collective goods, such as the environment, is also—or even more—evident on a global scale. In this sense, also, credit is due to the neo-Marxist Treadmill of Production adherents warning us not to be overoptimistic about the environmental motives and contributions of economic actors and dynamics per se. As a rule, the self-regulating economic actors have to be put under "pressure" first before they contribute to environmental improvements (leaving the few "win-win" situations aside). Political decisions, civil pressure, and citizen-consumer demand are decisive. But while they may arise in one corner of the world at a certain point in time, the economic "domain" has a strong role to play in articulating, communicating, strengthening, institutionalizing and extending (in time and place) these environmental reforms around the world by means of its own (market and monetary) "language," logic and rationality and its own "force."[4] Transnational industrial companies, global markets and trade, global information and communication networks and companies, and global economic institutions (such as the European Union, multilateral trade treaties such as NAFTA, investment banks like the World Bank, the Asian Development Bank, and the European Bank for Reconstruction and Development, and international financial institutions) play—or, rather, are beginning to play—a vital role in this dynamism. Moreover, developing regions are generally more deeply affected by the global markets and economic actors than by supra-national political institutions, although this varies according to each country's degree of integration in the global economy (consider Kenya) and dependence on—a few—transnational companies (consider Curaçao). The environment becomes to some extent institutionalized in the economic domain. And thus global economic institutions, rules and actors do less and less operate according to only economic principles and can no longer be understood in mere economic logics and terms.

However, and in the face of sometimes harsh criticism, I have to be clear about two points:

• These economic dynamics behind environmental reform cannot be understood as an established "fact" in all countries, or the majority of foreign investments or trade, nor as an evolutionary development that will "automatically" unfold. These dynamics can only be interpreted as developments taking place, a transformation in the making in the global economy that can be identified and which might very well develop on an increasing scale in the decades to come. But at the moment it is still a process in *status nascendi,* accompanied by power struggles, standstills and even regression. Various developments point toward an institutionalizing of the environment in the economic domain, giving these transformations a certain degree of permanence. But at the same time there is no fundamental reason or principle preventing this process of ongoing institutionalization from stagnation or reversal.

• The economy-mediated environmental innovations and transformations as they are developing now are significant and a major first step, but they are far from sufficient to achieve anything like a sustainable global economy in the end. Economic mechanisms, institutions and dynamics will always follow economic logics and rationalities, which implies they will always fall short in fully articulating environmental interests and pushing environmental reforms, if they are not constantly paralleled and propelled by environmental institutions and environmental movements. Neo-liberals who would have us believe that we can leave the environment to the economic institutions and actors are wrong. Besides, since economic interests are distributed unequally, any environmental reform brought about by economic players will display similar inequalities, making the results sometimes ambivalent.

The role of global economic dynamics in environmental reform, as well as the ambivalences involved, can be illustrated by the adoption of the ISO 14000 standards. The increasing need to have ISO 14000 standards in order to get access to certain international markets has triggered a drive for environmental harmonization. However, in their analysis of the global introduction of the ISO 14001 standard for environmental management systems, Krut and Gleckman (1998) show that this economic push for environmentally harmonized reform can also have its drawbacks. For one thing, a major part of the developing nations were excluded from the design process of the standard. Another drawback is that the existence of the standard disables countries from moving beyond compliance toward more stringent environmental goals. Both the reluctance of global firms to work toward such new environmental "standards" as well as the limits imposed on the possibilities of governments to move beyond the WTO regulations that sanction these ISO standards, can hamper progressive environmental developments. In a similar way the global regime of foreign direct investments has led to a "stuck in the mud" situation, as it fails to provide the incentives for nation-states to engage in a "race to the top" of the environmental standards.

Thus, in conclusion, environmental reforms induced and articulated by economic dynamics, institutions and actors do take place, and we may expect them to become increasingly important. In ecological modernization terminology: the environment is slowly becoming institutionalized in the economic domain. But this process will continue to be challenged and criticized for some time, with the traditional economic interests on the one side, and those who belittle the environmental gains and emphasize the related and unequally distributed social drawbacks, on the other.

Dialectics of Markets and Politics   Analyzing the role global economic actors and mechanisms can and to some extent already do play in environmental reform, it should be emphasized that these economic actors and mechanisms are not footloose, either in the political sense or in the geographical sense. First, markets and economic actors have always been and will remain phenomena that in the end are politically sanctioned. It is not only that—as so many social scientists rightly state—contemporary markets are organized and regulated by political systems and that they could not function as absolutely free markets these days. It is also that global companies and global markets are in the end dependent on a political legitimization of their products and production processes, and increasingly environmental controversies are part and parcel of this legitimacy question. This was so when they operated primarily on a national level and this is not fundamentally different at the global level, however flexible all forms of capital have become in moving around the world. Environmental groups and their global networks, international media, global political actors and institutions, and states intervene in the markets and condition the actions of global producers. Second, markets and global firms have to settle in geographical locations. This is evident when it comes to material operations in terms of production, distribution and consumption of capital and consumer goods, but no less true for the operations of monetary capital (Sassen 1994). Although geographical flexibility has vastly increased, "even in a globalizing world, all economic activities are geographically localized" (Dicken 1998, p. 10). And in these localities, the economic interactions are organized, designed and shaped by extra-economic logics such as the local social, cultural, political and physical conditions, even if they engage with actors on the other side of the world. So if, from an ecological modernization perspective, we emphasize the growing importance of market dynamics in global environmental reform, we have to be aware that "das Projekt der Marktwirtschaft war immer auch ein politisches Projekt—eng verbunden mit der Demokratie" ("The project of the market economy has always been a political project as well—closely connected to democracy") (Beck 1997, pp. 232–233). The global market economy and its representatives are under constant scrutiny for the legitimacy of their performance regarding, among other things, the maintenance of the sustenance base, exactly because they are not footloose.

How, then, should we make more use of and reflexively strengthen the environmental reform dynamics of global markets and economic actors? The suggestions put forward from various theoretical and practical or empirical perspectives are numerous. Some argue the case for a system of product biographies, coupled with a labeling system, and fines in case of irregularities. According to its advocates, such a system could run largely on self-regulation and self-control, since consumer demand would reorient global production and products. Others argue for extended product liability, making the producers liable for environmental damage from the cradle to the grave of a product (Lindqvist and Litt 1998), and beyond national boundaries. Some of the other suggestions heard are: a re-regionalization of markets (limiting transports, and energy consumption); the creation and protection of niche markets for ecologically sound products (Schot 1998); an expansion into tailor-made product-service combinations instead of continued mass production with high material intensity; transnational corporations' avoidance of certain markets in risk of becoming politicized by consumers and environmental movements (e.g., the market for products made by genetic engineering or recombinant technology[5]) (Beck 1997, pp. 234–245; Reinhardt 1999). Rather than listing all the options, these examples are meant to show that there is no shortage of ideas for using global markets and economic dynamics to attain global environmental reform. But political backing (in the broadest sense) is always needed to get markets and economic actors moving in a desirable direction, before market and economic actors can "take over" by articulating and institutionalizing the environment in their domain.

It remains a fact that the political drive to activate global markets and economic relations for environmental reform still comes mainly from the triad. This is something that is rooted in history, since the triad countries were the first to experience very severe environmental problems, as well as protests, and are currently less occupied with the basic economic needs or scarcity. For this reason, and owing to unequal power distributions, the environmental priorities and definitions of the triad are dominant in global economic institutions (and often also in multilateral environmental agreements), while developing countries often see their environmental priorities neglected. Furthermore, developing countries are at a disadvantage in initiatives to redesign economic institutions to incorporate environmental priorities (e.g., the WTO; see Sampson and Chambers 1999), or to conclude multilateral environmental agreements (e.g., the developments in the Framework Convention on Climate Change; see Gupta 1997). Re-negotiation of the Multilateral Agreement on Investment (either in the framework of the OECD or, more likely, in that of the WTO) and "greening" of the WTO (e.g., eco-labels, precautionary principle, article XX revisions) will be future cases where these environmental disparities between triad and non-triad will re-emerge (Opschoor 1999). Only strong political backing from beyond the political elites of the triad can ensure that these global economic institutions contribute to future environmental reform in a mode that is less "biased" toward the triad.

Global Civil Society

It often seems that those who are furthest removed from the actual practice of what has become known as the global civil society, are the first to emphasize the growing countervailing powers of the global environmental movement and the universality of environmental norms and principles, and the first to acknowledge the intensity of the pressures that the "civil society" has brought to bear on global capitalism. Among them are not only the captains of transnational industries such as Shell and General Electric, leaders of economic institutions such as the OECD (1997), or neo-liberal economic scholars such as Kenichi Ohmae and World Bank president James Wolfensohn,[6] but also (former) political world leaders from the triad such as German President Roman Herzog and US Vice-President Al Gore. All of them have emphasized the role of a globalizing civil society, tightly connected to the communication and information revolution, in achieving environmental reforms. Meanwhile, environmentalists, and the social scientists and political commentators closely linked to environmental movements, are much more cautious, ambivalent or even pessimistic as to the achievements of a global civil society in taming the treadmill of global capitalism. In most of their messages they continue to underline the dominant pattern of global capitalist developments, almost as if it had not been touched by the relentless efforts and pressures coming from environmental movements, "green" politicians, relatively marginal global environmental organizations such as the United Nations Environment Program and a diffuse and intangible global environmental consciousness. How to explain such contrasting evaluations?

In fact, there are several explanations. One of them is the political "game," in which the environmental movement creates itself an underdog position in order to be able to "beg" for the massive support it needs to beat the Goliath of global capitalism. By the same token, representatives of global capital overstate the strength of this movement, in order to suggest that sufficient countervailing power exists to balance global capital or even to point at the dangers of these powerful groups and ideas and legitimize a "green backlash" (Rowell 1996; Switzer 1997). To some extent it will also be caused by the view that the neighbor's grass is always greener, while the moss and ill weeds are most easily spotted in one's own backyard. From the perspective of transnational companies, Greenpeace must look like a powerful, well-organized and influential organization that manages to articulate environmental anxieties and consciousness into well coordinated campaigns that attract widespread media attention and increasingly force global economic players into retreat (and reform). The Brent Spar campaign is an illustration. From the inside, the perception is much more that of the difficulties of coordination between and within national groups, the failures in campaigns, the limited environmental results and the ambivalent relations with the media, or the tensions between the professional NGO offices and the large population of supporters and grassroots environmentalists.

Finally, it makes a difference whether one takes the economic triad as one's point of reference, the newly industrializing economies, or the developing countries of, for instance, sub-Saharan Africa. The newly industrializing countries in particular have witnessed a major fortification of the environmental movement and environmental consciousness under the recent conditions of globalization. The Latin American environmental movement, for example, has become much stronger during the last decade of the second millennium, and so have those in a number of newly industrializing economies in Asia (e.g., Thailand, Taiwan, the Philippines). Most African states and countries like Vietnam and China, however, have not spawned more than a rather scattered environmental movement, strong in some localities on specific issues (more than incidentally with Official Development Assistance), but rather powerless on a national level and poorly integrated in global networks. Nevertheless, even if civil society initiatives are still weak in African countries, the changing global power relations on environmental issues can have profound repercussions on transnational companies investing in the continent, as Shell has experienced in Nigeria and a consortium of oil multinationals and the World Bank are experiencing in planning—and, from October 2000 on, implementing—a major oil pipeline through Chad and Cameroon.

Global Environmentalism   The global civil society is not global in the sense that it has a global network of environmental NGOs covering every locality of the world. Nor does it have a common frame of reference similarly articulated in every corner of our planet. That will take some time, if not forever, to be accomplished. One obstacle is that any "shared" environmental frame of reference falls apart in different parts of the world. People's environmental priorities are different in different parts of the world (climate change versus clean water; nature conservation versus the "brown agenda") and definitions of environmental problems diversify as they are mediated by local backgrounds, history, and traditions. Environmental universalism is prevented by local factors articulating in heterogeneous cultural frameworks, as is widely acknowledged (chapter 2 above; Tomlinson 1999). But the most important cause behind the absence of a global frame of reference is the fact that the capacities and resources to articulate an environmental discourse in civil society are unequally distributed, especially—but not only—along the economic divides.

Here are the main reasons why we can still speak of global environmentalism:

• The ethics and principles of environmental behavior as regards the investment, production, and trade of transnational companies and investment banks are increasingly applied in a similar way to practices anywhere around the world.

• The potential to monitor environmental (mis)behavior of transnational corporations and institutions has moved far beyond the major centers of the global environmental movement in the triad.

• Environmental misbehavior and information are communicated around the world.

• Sanctions can transcend the boundaries of one state and are no longer limited to the localities of misbehavior. However, even though environmentalism has become global, transnational investors that are not strongly connected to the triad have less to "fear" from a global civil society. In Vietnam, for instance, American multinationals such as Nike are more vulnerable to environmental protests from the global civil society than regional investors from South Korea, Taiwan, or Indonesia.

The emergence of a global civil society with growing power to challenge the environmental destructiveness of global capitalism has made some of the major global economic players more aware of the need to move beyond mere compliance with formal political requirements laid down in laws and agreements. We can witness new forms of global environmental (sub)politics, arising especially in situations where nation-states are losing control on national and global developments, where scientific "proof" is no longer taken for granted but is increasingly seen as both an instrument of social interests and an object of social conflict, and where information and communication systems heighten the transparency of the worldwide actions of global economic actors. The controversies surrounding genetically modified organisms are of course a typical example where formal political requirements are overtaken by civil society politics. The representatives of global capitalism are finding it increasingly difficult to ignore civil society environmental protests and sensibilities, while formal environmental policies (both nationally and internationally) are "lagging behind." Transnational corporations are experiencing that they can less and less afford to restrict their environmental performance to compliance with formal political requirements. We can identify an increasing need, particularly for the visible multinationals from the triad, to justify their actions not only toward the states, multilateral agreements on environment, and conventional political actors, but also toward representatives of civil society, resulting in new forms of global environmental politics.[7] Major European food companies are developing various strategies to get into contact with (organized) consumers and concerned citizens, in order to become aware of their ideas and "sensibilities" in an early stage of product development. This does not result automatically in major environmental improvements, as many concerned citizens experience today, but it does give us a first glance of the potential stepping stones in the process along the way toward future global governance.

Toward Global Environmental Governance

Having outlined the central social dynamics at work in global environmental reform, the following question emerges: Can we use these insights to project a trajectory toward and a system of global environmental governance?

As I stated in chapter 2, globalization is a multi-dimensional phenomenon, which makes it impossible for us to identify one single cause or moving factor and one single outcome in terms of a future trajectory. The economic, political, cultural, social and ecological dimensions of globalization interfere and interact, heavily mediated by the revolution in information and communication technologies, and show both homogenizing and heterogenizing aspects. It would be naive—and a denial of the condition of reflexive modernization—to outline one likely and/or desirable system of global environmental governance here to confront the treadmill of global capitalism. But this study provides some insights that can be helpful in looking ahead toward any model or system of future global environmental governance.

First, any idea of "global governance from below" that is not matched by supranational governance institutions can no longer be taken seriously. Notions of decentralized forms of political organization in bioregions, empowering local communities for greater self-reliance, constructing local producer-consumer cycles, and the like, could still make sense in the 1970s. However, at the turn of the millennium, the age of globalization has radically restructured the institutional order, putting these kinds of governance alternatives in a different perspective. The dynamics of economic, political and cultural globalization are so massive, widespread and forceful that they can no longer be controlled by local structures and institutions alone. The choice sketched by Elliott (1998) between a radical local system of environmental governance and a reformist centralized one, is therefore false: radical is not linked to "globalization from below," centralized (or rather global) institutions are inevitable to control the treadmill of global capitalism. Moreover, local and global structures in world politics are increasingly linked in a non-hierarchical way.

Second, in the era of globalization, environmental reforms are no longer primarily managed or governed by conventional nation-states and their political institutions, individually or collectively. Any future system of global environmental governance will build upon the system of nation-states, but at the same time move beyond it. The "enmeshment of national political communities" (Held et al. 1999, p. 445) in regional and global processes affects the position of the nation-state with respect to subnational, national and global economic and social forces and actors. This is not to say that contemporary globalization processes can no longer be subjected to political decision making, steering and governance. Globalization is not beyond regulation and control and does not result in the "end of politics." Nor do I want to claim that the conventional units of political decision making and governance during high modernity, the nation-states, have now lost—or will soon lose—all sovereignty and autonomy and are on their way to becoming powerless agents floating on a stream of global developments triggered by markets and their dominant actors. The central insight from this study is rather that under conditions of globalization nation-states and national political actors are embedded in broader frameworks of governance and politics, and that these frameworks consist of multiple layers, from local to global, and multiple actors, from private firms to non-governmental interest groups. The Westphalian system of nation-states that has dominated domestic and international politics for such a long time is undergoing profound transformations. That means that international relations and politics will not be the only institutions of global governance. Wapner (1996), for instance, is right in underlining the importance of non-state civic forces and mechanisms in world politics: they will become stronger and gain greater autonomy. Although they will remain related to the nation-state system in some way, these new mechanisms of environmental governance are not unilaterally dependent on the states, as was the general interpretation two decades ago. Strengthening environmental governance is no longer equal to strengthening the nation-state(s).

Third, any powerful future system of global environmental governance will include greater "independence" or autonomy of environmental interests, institutions and arguments from their economic and political counterparts. Global environmental governance cannot remain restricted to the greening of existing economic institutions such as the IMF, the World Bank, and the WTO. Current and planned reforms show us that in these institutions greening usually means adding one criteria (environmental consequences) to the evaluation of existing, and the design of future economic projects, rules and practices. Although this can be beneficial in the short term and a major step forward compared to two decades ago (depending on how the greening is operationalized), such a greening strategy will never bring environmental interests on a par with economic interests. To give an example: expanding the scope of article XX of GATT or designing a new WTO agreement on MEAs would seem major environmental improvements. But they are only small steps if the final aim is that products, production processes and investments should be evaluated in terms of environmental objectives at the same level as they now are in terms of trade and investment liberalization objectives. Serious global environmental governance implies that the central rules and institutions that govern human activities world-wide are redesigned from an ecological rational point of view. In principle, this can be done within the framework of the dominant global institutions of today, integrating environmental interests with economic, security and other interests. But history "shows" that it is not easy to get the representatives of dominant value systems to accept previously subordinate values on an equal level. Especially since a kind of "supra-state" structure is lacking, radical environmental reform will require setting up environmental institutions on a regional or global scale, next to the major economic, military, and other institutions. These environmental institutions can of course take different forms, be issued with different mandates and have different relations to nation-states, depending on the outcome of concrete power struggles, among others.

Fourth, no matter how global environmental interests are articulated, they will contribute to the relocation of effective political power away from the (system of) "sovereign" nation-states. The development of a global environmental regime in terms of cultural norms and values regarding sustainability, has already contributed to a situation arising in international environmental politics in which a state can no longer simply justify its actions with an appeal to the principle of sovereignty alone. Environmental policies and politics of nation-states are no longer legitimate because the state is sovereign; they have to be legitimized on the basis of environmental arguments in a global arena.

Finally, any emerging system of global environmental governance will raise additional questions, beyond that of effective—that is ecologically rational—environmental reforms: questions of legitimacy, equity and democracy. In the heydays of national environmental governance, too, the procedures, processes, institutions and outcomes of environmental governance were subject to constant debate on such themes. But in an era of globalization, with global environmental governance emerging, something new and special is at stake. As long as the basic political and representational categories and institutions of the world remain based on nation-states (with the EU as a partial exception), while the most powerful forces shaping our world increasingly escape the boundaries and control of these units, we fall into a democracy and legitimacy vacuum.[8] And if a system of global governance consequently moves increasingly beyond the conventional political categories of the (system of) nation-states, then the conventional national solutions to questions of legitimacy and democracy become even less effective. This adds considerable weight to the need for democratic, legitimate and just global (environmental) governance structures. It is beyond the scope of this book to outline such new supra-national political and decision-making structures, institutions and dynamics that will need to be combined to form a future system of democratic global governance that is able to constrain global capitalism's devastating environmental impacts as well as achieve a number of other cherished social goals. Others (most notably Held 1995; Held et al. 1999; and the quotations in these studies) have provided some building blocks and outlines for global governance. Such new forms of environmental governance that surpass the conventional national and international political institutions, and with that the conventional notions of democracy, will not only have to be judged by their legitimacy and democratic quality; they can also be seen as contributing to the development of new modes of overall democratic governance on a global scale. Though we will not very likely witness the fading away of the system of nation-states in the near future, it would be a serious oversight not to take the consequences of the current changing character and role of the system of nation-states seriously.

The Ecological Modernization Theory in an Era of Globalization

Now that the central globalization mechanisms working toward environmental reform—both existing and in *status nascendi*—have been analyzed, and the consequences for future global governance assessed, I will finally return to our theoretical orientation. Continuing and extending the analysis and discussion of chapter 3, this section aims to summarize the consequences of this study for the Ecological Modernization Theory. For a relatively long time, the Ecological Modernization Theory focused both theoretically and empirically on the environmental reforms taking place in a number of industrialized, Western countries, even though the importance of supra-national and global dynamics for these new patterns of (national) environmental reform has been recognized (Weale 1992; Mol 1995). On the basis of the present study I am better able to refine the theory, seeking to answer such questions as: what are the consequences of the analysis of globalization-induced environmental deterioration and reform for the

Ecological Modernization Theory? Do these global "ecological modernization dynamics" in fact indicate that the ecological modernization theory becomes more uniformly valid for a much larger part of the world?

Meta-Theoretical Claims of the Ecological Modernization Theory

To develop my argument it is necessary to go back to the essence of the Ecological Modernization Theory. The basic, most fundamental, idea of the Ecological Modernization Theory has been formulated as the "emancipation," "differentiation" or growing independence of an ecological sphere and rationality with respect to the economic sphere and rationality, in particular (Mol 1995; Mol 1996; Spaargaren 1997). Elaborating on this relatively independent ecological logic and domain, the Ecological Modernization Theory has created a conceptual space to study contemporary institutions and social practices from a specifically ecological "point of view." This conceptual refinement was claimed to be bound to a specific time and space since it reflected actual developments in practices and institutions in industrialized society, where environmental ideas, practices and interests were clearly articulated and institutionalized from the 1970s on, with a major acceleration after the mid 1980s. Since the 1980s, social practices and institutional developments in the sphere of production and consumption have been "infected" with this emerging ecological rationality, resulting in major or minor changes—be it to a different extent, in different ways and at a different pace in the various industrialized societies (Mol 1999b). The conceptual innovation of the Ecological Modernization theory enabled us to analyze and understand these processes of transformation and reform.

It is my contention that, formulated at this meta-theoretical level, the attempt of the Ecological Modernization Theory to bring the environment (back) into social theory has proved to retain its relevance under conditions of globalization. The empirical evidence presented both in this volume and in numerous other publications, provides a sufficiently fertile ground to claim that the Ecological Modernization Theory is a valuable conceptual framework to gain an understanding of the ways in which environmental considerations and interests trigger changes in global institutions and social practices that are heavily infected by globalization. In this chapter I have summarized how the environment is articulated in these global institutions, and at the same time puts under pressure and changes the rules, procedures and functioning of these institutions. The system of nation-states, the global markets, the world wide economic and political institutions, the global civil society, are all "put to work" in greening global production and consumption processes; but at the same time all these institutions are transformed in the process of global environmental reform itself. In my view there is little foundation for the claim that emerging globalization processes have "bleached all the green" out of production and consumption processes, which would render any analysis from an ecological "point of view" point less. The institutionalization of the environment in social practices and institutions continues also under conditions of globalization and via globalization processes and dynamics, be it not in an evolutionary way of success upon success. And since it is the Western industrialized societies that are leading the way in creating, designing and governing global environmental institutions and in "determining" environmental-induced transformations in all kinds of social practices and institutions—as claimed in chapter 3 and empirically illustrated in chapters 7 and 8—this institutionalization of the environment is causing increasing homogenization rather than increasing heterogenization.

Environmental Glocalization

However, this "meta-theoretical homogenization" of ecological modernization, unequally determined by Western industrialized societies and their institutions and actors, but by no means controlled by them, converts into heterogeneous practices, trajectories and processes of environmental reform in the different countries and regions.

In chapter 3, I have outlined the European-based ecological modernization heuristics, which enabled us to understand national environmental reform dynamics in that corner of the triad. Having confronted these ecological modernization heuristics with the distinct countries and regions around the world, one critical conclusion seems to emerge straight away. In numerous countries and regions environmental reform—if it takes place at all[9]—only seems to "follow" some of these heuristics, and then often in a specific (national or regional) form and tempo. Sometimes civil society organizations play no significant role at all or its ideologies and strategies do not change according to these heuristics (as in Vietnam and China), sometimes political modernization processes take a different course or keep a different pace, sometimes (as in the United States) political modernization does little to explain environmental reform, and in other cases environmental considerations do not seem to become institutionalized to any significant extent in economic and market forces.

The background of these distinctions in ecological modernization is to be found in national institutional differences. These include state-market relations ranging from developmental states to predatory states, or from a "Rheinländisch" model to an Anglo-Saxon one (Evans 1995; Staute 1997); national policy styles (Richardson 1982; Vogel 1986; van Waarden 1995); regimes of accumulation (e.g. from extensive to intensive; see Lipietz 1987); national systems of innovation, with their national-specific network of institutions that initiate, import, modify, and diffuse new technologies (Nelson and Rosenberg 1993; Edquist 1997)[10]; and "national character," which Cohen (2000) ranks from weak to strong environmental consciousness and from numinous-aesthetic to rational-scientific epistemological commitment. These and some other aspects have been found to give ecological modernization a specific national or regional "flavor." And it is on this level that some of the notions and ideas contained in earlier contributions to the Ecological Modernization Theory, such as the presented heuristics, are in need of contextualization. Under conditions of globalization, processes of political modernization, the changing ideologies and strategies of environmental movements, shifting technological trajectories and economic internalization dynamics continue to be relevant as categories to look at in order to understand environmental reform in very different parts of the world. But this study makes us aware of two additions:

• On a global level—where globalization processes are "at work"—the nature and causal mechanisms of ecological modernization dynamics differ from those identified by ecological modernization theory on a national scale in European industrialized societies;

• Although the increasing importance of globalization processes and dynamics—and the articulation of environmental interests in it—adds to the relevance of ecological modernization for a growing number of regions and countries, the heuristics that "govern" environmental reform will always be "co-determined" by national and regional characteristics: environmental glocalization.

A Global Research Agenda

The geographical limitations of the Ecological Modernization Theory were identified at an early stage (Mol 1995, pp. 54–57; Spaargaren 1997). Critics of the ecological modernization theory have also used this argument to question the value of this theory as such in an increasingly globalized world. On the other hand, the supposed geographical limitations have also been challenged by scholars who have used ecological modernization perspectives in studies of environmental reforms in non-OECD countries (Frijns et al. 2000; Rinckevicius 2000; Sonnenfeld 2000). The conclusion of this volume is that both groups are partly right (and thus partly wrong).

Both I and the critics have been suspicious of ideas which seem to claim that environmental reform processes show universal forms, dynamics, and characteristics, in view of the fact that nations and regions differ and that environmental reform mechanisms vary accordingly, no matter how strongly such environmental reforms are triggered and influenced by global processes. Local refinements and contextualization of this theoretical framework, which until recently could rightfully be criticized for being too monolithic, too Eurocentric, would be most welcome (although that is certainly not what all critics have in mind—see Blühdorn 1999 and Pellow et al. 2000).

The other scholars are right in so far as they claim that under conditions of globalization, "global" ecological modernization—to a large extent defined by triad countries—has an universalizing effect on the way in which countries experience and design environmental reforms. Defined in a not too strict way, ecological modernization was perhaps more specifically Eurocentric and unique 10 years ago than it is at the start of a new century marked by globalization. Studying and defining regional (or national) "variations" or "styles" of ecological modernization seems to me a promising in-between course for understanding and interpreting innovations and achievements in environmental reform under conditions of globalization as well as for outlining future trajectories for environmental transformation. That will enable us to see how far the commonalities of Ecological Modernization reach, where the specifics of the regional variation start, and how the two change in time.

Back in Balance: Back to Seattle

Preceding the Conference of the Parties (COP-5) on the Biosafety Protocol (of the Biodiversity Convention) in Nairobi May 2000, rural and environmental NGOs have strongly protested against new technological developments by the so-called Gene Giants, the transnational seed and agrochemical corporations. Genetic seed sterilization, popularly called

"Terminator," engineers crops to kill their own seeds in the second generation, making it impossible for farmers to save and replant seeds. More than 1.4 billion people, primarily the poorer farmers in the South, depend on farm-saved seeds. The related trait-specific genetic use restriction technologies (T-specific GURTs) enable companies to let seeds perform only with a specific chemical, pesticide or fertilizer. The success of these products would reinforce chemical dependencies, and thus use. Both can have a major impact on biodiversity, food security and costs for farmers. The year 1999 saw the acceptance of seven new Terminator patents and several field trials of GURTs.[11]

The demand for a ban on these new technologies in accordance with the precautionary principle in the Biosafety Protocol is a test case for the post-Seattle era. But that is not the main reason for quoting this example. This example—out of many others—makes us aware that the emphasis on global environmental reforms in especially the second part of this volume should not make us ignore two major continuities that are still valid: First, the institutionalization of environmental reforms is not the "natural" rule yet, but only too frequently merely the result of action after the fact, in response to plans and patterns of economic globalization that threaten the environment. Second, environmental NGOs and other civil society actors continue to play a crucial role—also on a global level—in setting into motion major environmental reforms, also at a time when transnational companies and organizations claim to have understood and to behave according to green rationalities. The protests at Seattle were unique for their broad scope, the wide variety and massive numbers of protesters and their radical outlook. But similar global events setting the environmental agenda of institutional change can be expected. These calls for global institutional change are just as much a product of the globalization era as the increases in FDI or the push for trade and investment liberalization.

Notes

Chapter 1

1. There were a few considerably earlier introductions. World-System theorists claim that globalization is an "old" World-System Theory concept that has become a buzzword for the present phase of capitalism. See Roberts and Grimes 1997.

2. Hirst and Thompson are certainly not alone; see Gordon 1988 and Glyn and Sutcliffe 1992.

3. This position is still defended in basically the same way by writers such as Hobsbawm (1994), albeit against the background of increasing global interdependence.

4. Hoogvelt (1997, p. 116) claims that, in the confusing and spreading discourse on globalization, sociologists—rather than economists or international relations scholars—have been at the forefront in trying to give globalization a consistent theoretical status. Although this is true in the main, with the notable exceptions of Dicken (1998) and Ruigrok and van Tulder (1995), it is not for this reason that I concentrate on sociological studies and pay only limited attention to the contributions of international relations experts and economists. The focus on sociological perspectives is in line with the central objective of this book.

5. As Nederveen Pieterse (1997) stresses, the hypothesis of delinking or decoupling from world capitalism is still present in the globalization debate, with arguments running along three lines: various pleas for local economic development, arguments for a return to national Keynesian management and "new protectionism," and anti-development views. However, neither of these three lines is as radical and drastic as the proposals advocated in the 1970s by scholars sympathetic to the Dependencia school of thought.

6. Recently the OECD entered the discussion on globalization and the environment primarily from an economic perspective with several studies and conferences (OECD 1997a,b; Gentry 1999; Goldenman 1999). In more concrete political terms, the OECD has contributed to the debate with its proposal for a multilateral agreement on investment. Other international economic organizations, including the WTO and the EU, have gone through similar developments.

7. One field of study in which globalization has been linked to the environment is the one that concerns globalization of agriculture and food chains (Whatmore 1994; Bonnano et al. 1994; McMichael 1996b). However, most of the studies in that field emphasize the devastating consequences of globalized food production for the environment, for food quality, and for related human health issues, rather than for environmental reform.

8. Although he only marginally and peripherally touches upon processes of environmental devastation, perception and reform, Manuel Castells's (1996, 1997a, 1997b) work on globalization and the network society is of special relevance for the present analysis, as it shares some of the basic starting points and characteristics of the ecological modernization theory. There are of course various more specific—and thus limited—studies that concentrate on, for instance, the emergence of a global environmental movement, environmental international relations and regimes, or global trade and/or investment and the environment. I will make extensive use of these studies in the coming chapters. However, these studies do not combine the various forms and views of globalization with (changing) environmental deterioration, environmental perceptions and environmental reform. They do not therefore succeed in giving an overall perspective. The 1992 United Nations Conference on Environment and Development (UNCED) is often mentioned as a positive example of global environmental management and reform.

9. Redclift (1996), however, is not really concerned with globalization as such, but basically deals with the distribution of consumption between the "North" (i.e., the wealthy industrialized countries) and the "South" (i.e., the developing countries). As I noted earlier (Spaargaren and Mol 1995), only a few of the contributions in Redclift and Benton 1994 deal with globalization.

10. The number of contributions to this idea seems endless. Ohmae 1995 is one of those most frequently cited.

Chapter 2

1. This chapter examines the academic debates surrounding these questions within the fields of sociology, international relations, and economic history, focusing on the transformation and continuity of the modern order. In order to elaborate the arguments in this book and achieve a better understanding of the changing conditions for environmental protection and regulation, a detailed understanding of our changing modern order—and especially the emergence of globalization—is essential. Readers who are more interested in the actual relationship between globalization and the environment may find the following three sections too theoretical and unfocused and may prefer to continue with the section that deals with the more substantial aspects of globalization in the economic, cultural, and political spheres. Globalization specialists are likely to be largely familiar with the literature and debates dealt with in this chapter and may wish to skip it altogether.

2. To some extent this parallels the distinction made by Held et al. (1999, pp. 3–10) among the hyperglobalizers, who celebrate globalization as a natural phenomenon that brings economic fruits to the world and the end of the nation-state; the skep-

tics, who deny that globalization is anything new in either economic or political terms; and the transformationalists, who see globalization as a major force behind the rapid social, political, and economic changes that are reshaping the modern world in an unprecedented way, although the direction is uncertain and its fruits remain to be seen.

3. Except perhaps the fall of the Berlin Wall and the disappearance of state socialism in Europe, although in retrospect it appears that these things happened less suddenly than most witnesses acknowledged in 1989.

4. Most adherents to a discontinuist interpretation of globalization converge on this time frame. See Giddens 1990; Amin and Thrift 1994; Group of Lisbon 1995; McMichael 1996b; Castells 1996, 1997a,b; Dicken 1998; Held et al. 1999.

5. Among the first extended and interesting attempts to link World-System Theory to environmental deterioration (and, to a much lesser extent, to reform) are Sonnenfeld 1999, Goldfrank et al. 1999, and the special issue of the *Journal for World-Systems Research* (3, no. 3, autumn 1997).

6. Recently these "older" monolithic theories have incorporated a multiplicity of factors that contribute to globalization. Adherents to the World-System Theory school of thought, for instance, have dealt with the more cultural and ideological factors in globalization processes and tried to incorporate these into their theory (Wallerstein 1990).

7. For World-System Theory see Roberts and Grimes 1997 and Frankgold et al. 1999. For the international relations school of thought see Haas, Keohane, and Levy 1993; Keohane and Levy 1996; Young 1997a, 1999.

8. Robertson himself (1992, pp. 138–145) thinks of Giddens's work on globalization as a rather tardy contribution that adds little to the existing theories.

9. These concepts and terms will all be elaborated in this study.

10. Bearing this in mind, one can understand why Robertson (1992, p. 145) criticizes Giddens's work on globalization as an "overly abstracted version of the convergence thesis."

11. Although all four institutional clusters are relevant for understanding the emergence and persistence of environmental threats in relation to globalization, Giddens considers the most direct link to run via industrialism and the global division of labor. See chapter 3 below.

12. The emphasis on flows rather than on national institutions occurs in many recent contributions to globalization theories. Lash and Urry (1994), for instance, identify the present age as one of the disorganization of capitalism. According to them it should be understood not so much as the successor of organized capitalism but as a global system that is systematically disorganized, and in which the logic of organizations, especially the nation-state, is replaced by the logic of flows.

13. These themes and concepts are also often reinterpreted and transformed. For instance, the notion of the information society was originally closely connected to the emergence of the service sector with respect to industry and especially agriculture (Bell 1976). More recently, the relevance of information and telecommunication technologies for any specific economic sector has been downplayed. Now, these

technologies are instead linked to major transformations in all production, distribution and consumption activities (Castells 1996).

14. During the colonial period up to 1960, the so-called Third World received half of total direct investment flows. This percentage has declined to 25% in 1974 and to only 16.9% in 1988–89. More than half of the latter amount went to South and Southeast Asia (Hoogvelt 1997, p. 77). In the 1990s this percentage of total direct investment in South Asia increased slightly.

15. Amin (1997) strongly criticizes Hirst and Thompson for their quantitative evaluation of globalization and neglect of qualitative shifts, according to which "anything between globalization as an international trading system and globalization as the end of the territorial state" is disregarded (Amin 1997, p. 126). Nederveen Pieterse (1997, pp. 372–373) joins this critique, underlining that Hirst and Thompson limit globalization to economic globalization, that they are more concerned with a truly globalized economy than the process of globalization, and that their time frames are ill chosen.

16. See, e.g., Gordon 1988; Patel and Pavitt 1991, 1994.

17. Some data on capital flows can be found in Scholtens 1998. Total official capital flows decreased from about US $60 billion (1990) to US $36 billion (1996–97), mainly because of a drop in bilateral ODA (Official Development Assistance). Private capital flows skyrocketed from US $45 billion in 1990 to almost US $250 billion in 1997, with FDI accounting for around 45–50% of these total private flows.

18. Castells (1996, p. 101) provides average FDI data for 1989–1991, the OECD (1998) for 1990–1997, and UNCTAD (2000) for 1998–1999, all showing the continuing global dominance of the triad in world FDI. The US and the UK have been firmly among the top of world FDI outflow countries for several years, although FDI from Japan has been increasing rapidly (Dicken 1998; UNCTAD 2000).

19. Hong Kong and China, Taiwan, Singapore, China, South Korea, Malaysia, and Brazil were among the top 25 world FDI outflow countries in the period 1990–1996 (OECD 1998). Four-fifths of the FDI outflow from developing countries comes from these seven countries, six of which are newly industrializing Asian economies (Dicken 1998, p. 45).

20. The "tigers" are countries that showed high economic growth in the last quarter of the twentieth century. FDI inflow to developing countries had fallen to 26% of world FDI in 1995. Africa's share of the developing countries' inflow was only 8.3% in 1995, while Latin America's share also dropped to 33%. The Asian share increased from 21% in 1975 to 58% in 1995 (Dicken 1998, pp. 47–48) The close relationship between the top FDI inflow and the top FDI outflow countries reflects the concentration in economic globalization, even in non-triad countries. See also UNCTAD 2000.

21. FDI in services increased sharply from the 1980s on, while FDI in extractive industries—historically the most important—has become less significant (although by no means negligible; consider the petroleum sector). Manufacturing had the bulk of FDI until at least the mid 1980s.

22. There is no need to say that this core and its antagonistic periphery are no longer geographical entities. The core-periphery relationship cuts across national and geographical boundaries, bringing in the core segments of the Third World and relegating groups and sectors of the traditional geographical core to peripheral status.

23. In providing evidence of the growing importance of non-state players, one of the leading new institutionalists, Oran Young (1997a), makes an analytical categorization of two kind of regimes: international regimes, where states are the major constituents, these being the central concern of modern institutionalists; and transnational regimes, which are institutional systems whose members are non-state players within global civil society. McCormick (1999) makes a similar distinction between IGOs (Intergovernmental Organizations) and INGOs (International Non-Governmental Organizations).

24. Smith (1993) and Hovden (1999) have criticized regime theorists for taking the present global institutional order as given rather than looking for political renewal; for neglecting the contribution from "below"; and for focusing on cooperation rather than on conflict, unequal distribution and power. According to these authors "an [international relations] discipline characterized by the dominance of regime theory and realism is poorly suited to study this crucial aspect of the contemporary environmental crisis" (Hovden 1999, p. 67).

25. A world polity was originally defined as a "system of rules legitimating the extension and expansion of authority of rationalized nation-states to control and act on behalf of their populations" (Meyer 1987, p. 69). Boli and Thomas (1997) put it in a broader perspective by focusing on both political and cultural institutions and by taking a transnational perspective rather than one within the framework of the system of nation-states.

26. In his analysis of the undermining of the sovereignty of nation-states Held makes a distinction between internal sovereignty, centering on the government as the final and absolute authority that exercises supreme command over a particular society, and external sovereignty, a notion which underlines that there is no final and absolute authority above and beyond the state.

27. Especially between the members of the international business class, state bureaucrats and members of international organizations.

28. Held includes environmental problems as one of his points.

29. Several authors, including Castells (1997a), have pointed to the emergence of local and regional governments as significant actors in the global arena. Institutional evidence is found in the EU, where in the late 1990s a council on regional government was established, given influence to sub-national (regional) authorities in supranational policy and decision making.

Chapter 3

1. "Factor x" means x times less use of energy and material per unit of GDP.

2. Countries as diverse as Russia, the United States, and most developed European states almost simultaneously witnessed the creation of nature reserves and parks

and the protection of endangered birds on the eve of the twentieth century. This provides evidence of a common cultural attitude toward nature among the elite, and an international exchange of ideas and strategies for nature protection. It paralleled the acceleration in the process of internationalization, as stressed by globalization theorists such as Robertson. Needless to say, the specific forms and strategies of nature protection differed among countries owing to specific national circumstances (van der Windt 1995, pp. 30–44).

3. Reopening this debate, Goldblatt (1996, pp. 14–51) criticized Giddens for restricting environmental degradation to the institutional cluster of industrialism and the international division of labor only, neglecting the environmental relevance of the capitalist dimension.

4. In terms of attention paid to the environment in daily newspapers and opinion polls, the heyday of this upsurge lasted until early 1993 in most Western countries.

5. For mixed evaluations and critiques of the idea of ecological modernization as the common denominator of environmental reform processes, see Hannigan 1995; Christoff 1996; Blowers 1997; Leroy and van Tatenhove 2000; Blühdorn 2000; Pellow et al. 2000; Pepper 1999; Schnaiberg et al. 2001.

6. "High consequence risks" are defined as risks that, although remote from control by individual agents, threaten the existence of millions of people and indeed humanity as a whole. The principle examples given by Giddens (1990, 1991) are Chernobyl, ozone-layer depletion, and the greenhouse effect.

7. See also the recently established *Journal of Industrial Ecology*.

8. These theoretical traditions include system-theoretical analyses by Huber (1985, 1991), more institutional analyses by Skou Andersen (1994) and Mol (1995), and discourse analyses by Hajer (1995) and Weale (1992).

9. Murphy and Bendell (1997, p. 63), for instance, summarize ecological modernization or ecomodernism as "the perspective that treats the environment as another technological problem to be overcome in the pursuit of progress." "To the ecomodernist," they write, "pollution is an economic opportunity for prevention and cleanup technologies and certainly not an indication of fundamental problems with the current economic system." This summary and implicit criticism would have been better in the late 1980s than at the turn of the millennium as it points to some of the already often-quoted shortcomings in the earlier contributions to the Ecological Modernization Theory which have been incorporated effectively in more recent writings.

10. Maarten Hajer, in particular, emphasized the increasing dominance of ecological modernization as new central story line in environmental ideologies and politics from the mid 1980s on. See Hajer 1995, 1996.

11. As Beck explains (1994a, p. 22), "sub-politics is distinguished from 'politics', first in that agents outside the political or corporatist system are allowed to appear on the stage of social design . . . , and second, in that not only social and collective agents but individuals as well compete with the latter and each other for the emerging shaping power of the political."

12. See Mol and Spaargaren 2000. This growing importance of the Ecological Modernization Theory is acknowledged even by its critics, who often do not chal-

lenge the analytical and descriptive qualities of this theory for West European societies but rather its normative undertones. Though contemporary environmental policies and reforms may indeed "be based on" or reflect ideas of ecological modernization, they should also be criticized for that, as such attempts to solve the environmental crisis suffer from various problems, according to these critics.

13. At this point I put the differences within countries between brackets, although these certainly exist and are relevant (e.g., between the central urban centers and the peripheral rural areas, or between large transnational corporations and local medium-size and small enterprises).

14. Arguing along similar lines, I have on various occasions questioned the Netherlands Ministry for the Environment for neglecting the institutional presumptions underlying its environmental policy model in its attempt to "export" this model to countries as diverse as Latvia, Hungary, Surinam, the Netherlands Antilles, and Russia, and—perhaps more successfully—to Canada, New Zealand, and several US states (New Jersey, Minnesota, Oregon, California, Pennsylvania).

15. The Ecological Modernization Theory as a normative program will have more practical relevance for Poland, Hungary, and the Czech Republic (also the most likely first candidates for EU membership) and less for Bulgaria, Albania, and several CIS republics. See also Huber 1993. Rickevicius (1998, 2000) and Gille (2000) have analyzed the value of ecological modernization ideas in more depth for, respectively, Lithuania and Hungary.

16. Wallace (1996, pp. 58–60) summarizes an alternative to ecological modernization in what he calls the "radical green agenda" (as opposed to sustainable industrialism).

17. The Eurocentric character of the Ecological Modernization Theory makes its relevance for the United States and other triad countries also subject to debate (Pellow et al. 2000; Schnaiberg et al. 2001; Mol 2000b; Buttel 2000b).

18. In *The Refinement of Production* (Mol 1995, pp. 48–50 and 397–401) I paid more attention to these two dimensions of social theories and to the question of their interrelatedness under conditions of reflexive modernization.

19. This is, of course, a much stronger version of environmental (neo-) colonialism than the one put forward by Agarwal and Narain (1991).

20. This line of reasoning runs largely parallel to the "tragedy of the commons," in which the breakdown of the traditional structures supporting these commons (by the emerging capitalist economy) resulted in their deterioration, only to be stopped—at least partially—with the introduction of modern institutions of environmental protection (Mol 1995; chapter 3). Now global capitalism is doing the same on an international scale, according to these scholars. I would argue that we are thus in need of new structures for environmental protection, rather than trying to safeguard the old, conventional ones.

21. As I will argue throughout this book, these tendencies of what some might call homogenization or even "Westernization" of environmental reform are already taking place in these countries. This occurs through international environmental negotiations, coordination and assistance and scientific-technological exchange of ideas, experiences, and hardware. In addition, global firms operating on a world market

and having learned the lesson of Bhopal help to spread this movement, as do international environmental NGOs (e.g. Friends of the Earth International) that stimulate their members all over the world to apply the idea of "environmental (utilization) space" in developing alternative environmental reform models for their home countries.

Chapter 4

1. "Apocalypse-blindness" refers to a modern society whose institutions have no "receptors" for sensing apocalyptic dangers at an early stage of progression (Beck 1994b, p. 180). Modern society may therefore be confronted by apocalyptic crises rather suddenly.

2. Consider Beck's notion of the boomerang curve.

3. Having reviewed the literature, however, I will conclude in chapter 7 that there is little evidence to support the thesis that systematic international relocation of industrial production is caused by differences in environmental regimes.

4. This has been slowly changing. See, e.g., Adeola 2000.

5. This basically restates the state failure theories of the 1980s posited by Jänicke (1986).

6. Clark (1996) also mentions the existence of a group of authors who assert that there is a "race to the top" in international environmental politics, driven by economic incentives and market forces. Clark himself cautiously moves toward this last school of thought in his analysis of the case of Australian transnational mineral companies, although he acknowledges the specificity of his case study.

7. As Redclift (1996, pp. 34–35) points out in relation to environmental threats, "global sovereignty is still largely a rhetorical device."

8. The boycott of Shell products in various European countries initiated by Greenpeace in order to prevent Shell from dumping the Brent Spar oil platform in the Atlantic serves as the most important example of "world sub-politics" in Beck's (1996) analysis.

9. Sometimes advancements in global governance go together with a fortification of the role of the state. Goldenman (1999) reports a growing consensus that nation-states do have certain responsibilities if commercial activities originating from their territory cause problems in another country's territory. For example, the Basel Convention required that shipments of waste be returned to their home countries if the waste could not be handled safely in the country of destination. The home country's responsibility for TNC activities causing environmental damage abroad is not clear, however.

Chapter 5

1. This is relevant in regard to various theoretical contributions to globalization, such as the World-Systems approach: "This is one point at which early analyses of

relations between the world system and environment can be furthered. For transnational corporations and core states are not only vehicles of environmental destruction . . . , they may also be agents of environmental advancement." (Sonnenfeld 1999). The recent World-System volume by Goldfrank et al. (1999) also fails to present the other side of the argument.

2. Grant (1993, p. 61) defines a stateless company as a firm whose ownership, board of directors, and senior management are internationalized. Such a firm operates globally; it is no longer based in one country, nor is it loyal to one country.

3. The transparency of the WTO is still rather limited. Consider, e.g., its Dispute Resolution System (Davey 1999). Environmentalists see the undemocratic and uncontrolled dispute settlement system as one of the main reasons for resisting the WTO (Wallach and Sforza 1999).

4. French (1998) writes that the US Overseas Private Investment Corporation and the US Export-Import Bank are both starting to strengthen their environmental guidelines. With respect to the Three Gorges Dam project in the Yangtze basin, the US Export-Import Bank announced that its environmental guidelines made it impossible to provide export credit support to companies such as the heavy equipment manufacturer Caterpillar. However, the bank's equivalents in other triad countries stepped into the breach, demonstrating the difficulties in "greening" national export credit and investment promotion agencies.

5. See Jänicke 1993 for the national innovations in political institutions. See Young 1997 for global political and regime innovations.

6. These changing models of environmental governance at the national level are, of course, not caused by globalization processes alone. Within the field of environmental policy making, new governance models have come up, partly in response to the limited success of conventional command-and-control modes of governance and partly in response to movements toward and demands for deregulation (Mol, Lauber, and Liefferink 2000).

7. Marsden et al. 1993, Bonnano et al. 1994, and McMichael 1996b are three ground-breaking studies.

8. There are contrasting examples of the environmental and health risks of locally produced and consumed food products. These consequences are due to inadequate monitoring, quality control, and technological innovation, and also to the limited advantages of economies of scale.

9. Five major categories of supranational institutional systems are sometimes distinguished: informational (gathering, analyzing, and disseminating data), normative (defining standards, international principles, and goals), operational (overseeing implementation and administering financial and technical assistance), rule creating (enacting formal, binding treaties), and rule supervisory (monitoring compliance) (Jacobson 1984, pp. 88–90).

10. Rowlands (1995), for instance, argues that the UNCED has reinforced the sovereign rights of states and rejected any supranational authority on monitoring, reporting, and inspection, in spite of the quest of some participants in the UNCED process for both "sub-politics" and supra-national politics. As will become clear in

chapter 6, the EU is also still struggling with supranational authority on monitoring and inspection.

11. See the various contributions in Young 1997a and in Levy and Young 1996 that generally concentrate on what Young (1997b) calls "piecemeal or issue-specific arrangements that may or may not be legally binding, that may or may not assign some role to the United Nations and its specialized agencies."

12. Proposals such as those of Harris (1992) suffer from some of the weaknesses and limitations that according to Nederveen Pieterse (1997) are often found in constructive approaches to global regulatory and institutional reform. It is not clear how these proposals relate to ongoing transformations. They often reflect weaknesses in institutional analysis. They often pay little attention to the relationship between these global reforms and reforms at local and national levels. They do not satisfactorily combine diverse interests and perspectives that could promote new institutional structures.

13. Redclift (1996), however, notes that the proportion of natural resources and primary food products to the total exports of Third World Countries has decreased considerably, from 50% in 1965 to 20% in 1986. In the same period exports of manufactured goods increased from 25% to 50% of total exports. Dicken (1998) reports a similar trend.

14. But often also in close coalition with unions, consumer organizations, parliamentarians, and others (Goldenman 1999).

15. This notion comes close to Sabatier's concept of advocacy coalitions, although Hajer (1995, pp. 68–72) notes some important differences.

16. Elsewhere (Mol 2000) I have elaborated on the shifts in the main ideologies, positions, and coalitions in which the environmental movements in the triad are engaged in an era of ecological modernization.

17. The Group of Lisbon (1995) rightly observes that the unofficial parallel conference of NGOs to the UNCED also demonstrated that the "global civil society" is still highly fragmented, uncoordinated, and internally divided along several lines: North vs. South, environmentalist vs. developers, reformists vs. "revolutionaries," those starting from local interests vs. those arguing from a global analysis, etatists vs. defenders of local autonomy, and so on.

Chapter 6

1. Since the establishment of the EEC, in 1957, the EU and its predecessors have not always been successful in this continuing drive for economic integration. Various periods of stagnation of the integration process can be distinguished, usually caused by large member states that were hesitant to continue the integration process and suffer the loss of sovereignty that was often believed to be the consequence. Especially since 1984, external economic factors have given a new impetus to the economic integration process.

2. The subsidiarity principle states that issues should be regulated at the lowest possible governance level.

3. Europeanization has, of course, also been questioned from other perspectives and motives. There even seem to be fractions among the Euro-skeptics that oppose the integration process because it forces an excessively progressive environmental regime upon member states, preventing the free market from doing its work.

4. In June 2000, Joska Fisher, Germany's Minister for Foreign Affairs in a coalition government of Social Democrats and Greens, called for developing the EU in the direction of a European federal state.

5. EU Directives must be translated into national law; EU Regulations are directly legally binding in the member states.

6. Besides the traditional (legal) enforcement routes that focus on formal implementation of directives and regulations into national law, the European Environment Agency was originally meant to facilitate and strengthen the monitoring and implementation of EU environmental policies in the member states. Its enforcement power has been severely reduced compared to the original proposals of 1989. Information gathering, information dissemination, and technical assistance are now its main (in fact its only) tasks (Axelrod 1997).

7. With the accession of some of the Central and East European states to the EU, this argument may well resurface, although the requirements for accession include compliance with the existing level of environmental policies and standards. Nevertheless, their accession will once again change the existing balance between environmental front-runners and laggards, as it did with the accession of Finland, Sweden, and Austria. Andersen and Liefferink (1997) show that the differences in environmental priorities and interests between the "green" EU member states prevent them from forming a solid environmental coalition. The same will be true for the laggards after the accession of Central and East European countries.

8. Despite the fact that several authors (e.g., Porter and van der Linde 1995a,b; Jänicke 1997) have pointed to a mutually reinforcing effect of stringent environmental policies and economic development, this hardly ever results in "going alone" strategies of stringent environmental reform in the European Union. The frequently observed empirical correlation between green front-runners and economic strong countries is not seen as a strong causal relationship by the political and economic elites.

9. For example: the not very democratic structure of the EU, in which the European Parliament (often considered to be greener than the other institutions) has only little influence in decision making; the unequal allocation of budgets and resources favoring the more productive sectors, most prominently the agricultural sector; the limited integration of the environment in other policy fields; a number of institutional characteristics; etc.

10. This uniformity of environmental standards in the EU may also affect foreign producers wishing to enter the growing EU market. The larger the market, the more rewarding (or even necessary) it is to set environmental production and product standards at the level of this market, and this dynamic affects producers in developing countries as well, especially those exporting to global markets (Sonnenfeld 1998).

11. The World Development Report 1997 (World Bank 1997a) indicates that the GNP per capita of Mexico is at least 6 times smaller than that of Canada and the US, and that its average annual growth over the period 1985–1995 was 0.1% as compared to 0.4% for Canada and 1.3% for the US. Relative energy use and CO_2 emission per capita run almost parallel to GNP per capita.

12. Some (e.g., Esty 1999) interpret the 1989 announcement of the Bush administration to start negotiations on NAFTA and the 1991 "tuna-dolphin" case under the GATT regulations as the start of the trade-environment debate, although this seems a rather American point of view. Some of the environmental discussions in the EU can also be seen as part of the trade-environment debate.

13. Charnovitz (1994, p. 53) points out that the "bottom line is that NAFTA has no positive provisions (i.e. telling governments what to do) on the environment." "What it has," Charnovitz continues, "is one negative exhortation. More significant is what is largely absent from the NAFTA, i.e. negative disciplines (telling governments what not to do) on environmental standards."

14. The sharp rift in the US environmental movement continued until the last phase of negotiations on the environmental side agreement (Esty 1994; Vogel 1997; Hogenboom 1998).

15. Although environmental NGOs wanted the plan to be incorporated in the NAFTA they did not succeed. Jointly, Mexican and US NGOs managed, however, to push for considerable revisions. The final plan (February 4, 1992) was still criticized for its limited focus on urban and industrial environmental problems, its failure to create a new bi-national agency for environmental protection, and limits on Mexican NGOs getting access to policy documents and information on environmental hazards.

16. An innovation in the process toward the NAFTA was the environmental review of NAFTA, covering a large range of potentially negative environmental effects of NAFTA and providing specific policy recommendations to minimize or overcome these dangers. The review was criticized by environmental NGOs, among others, for being incomplete (Esty 1994; Hogenboom 1998).

17. This is generally seen as a direct response to the famous tuna-dolphin case of GATT and the WTO, in which a GATT dispute settlement panel recommended against the US imposition of a trade ban against Mexican tuna to support US dolphin protection efforts (Esty 1994; Vogel 1997). This 1991 ruling—never formally adopted by GATT—put trade obligations on a higher plane than environmental considerations and aroused strong public interest and debate on the environmental consequences of the NAFTA negotiations, especially since it involved the same two nations.

18. For a more detailed analysis, see Johnson and Beaulieu 1996. Those authors argue that the NAFTA treaty—and especially its pollution haven clause—does not have sufficient legal and enforceable sanctions to prevent or revert downward regulatory pressure, should it occur. Similarly, they claim that the idea of upward harmonization of environmental norms and standards (chapters 7 and 9 below) is largely unenforceable. But that does not mean that both are not followed or have policy relevance in the implementation process of NAFTA.

19. The Mexican executive director of CEC was forced to resign early 1998 as his environmental advocacy was believed to be disruptive to US and Mexican national environmental reform programs (Esty 1999).

20. The equal distribution of the costs of the CEC (US $9 million annually) among the three parties is an additional concern for Mexico. The US $3 million represents a major share of the Mexico's annual budget for environmental protection.

21. The Web site of the Canadian Environmental Law Association (http://www.web.net/cela/sbnn.htm) lists other cases still in progress where private companies have filed claims against NAFTA states for limiting import or exports for environmental reasons (e.g., Sun Belt of Santa Barbara vs. Canada for preventing exports of billions of liters of fresh water, Metalclad (USA) vs. Mexico for refusing to approve a waste dump, USA Waste vs. Mexico for damages resulting from a discontinued concession for street cleaning and a permission to open a landfill in Acapulco, and DESONA vs. Mexico).

22. Discussions on the extension of NAFTA into a Free Trade Area of the Americas (FTAA) based on the NAFTA provisions are resisted by opponents of free trade as well as by those, especially within the US but also in Mexico, who are opposed to environmental and labor side agreements being attached to such a new trade record (Gershman 1998; Honey and Barry 1996). Esty (1999, p. 199) argues that the proposal put forward in the "fast track" legislation to extend NAFTA toward FTAA does not contain any environmental provision and as such represents a major step back. Discussions on the environmental consequences of the accession of new member states to NAFTA, Chile being the first in line, to some extent parallel those on the enlargement of the EU with Central and East European states, although the environmental balance is more positive in the EU case: CEE accession countries have to fulfill all environmental EU directives.

23. Such similarities are, of course, stronger in the European region and between the US and Canada. Mexico seems a separate case in terms of cultural background and economic development.

24. Castells (1996, p. 109) mentions the existence of at least five powerful economic networks in the region: Japanese corporations, Korean corporations, American multinational corporations, the network of overseas Chinese capital (the China Circle), and Chinese provincial and local governments (with their diverse industrial and financial interests).

25. Evans (1995, pp. 47–50) suggests that the origin of the concept of developmental states goes back to Johnson (1982) and White and Wade (1984). Following Evans, Castells (1997b, pp. 270–271) gives the following definition of a developmental state: "A state is developmental when it establishes as its principle of legitimacy its ability to promote and sustain development, understanding by development the combination of steady high rates of economic growth and structural change in the productive system, both domestically and in its relationship to the international economy."

26. Maull (1992, p. 357) observes that, in contrast to European countries, Japan has long been sheltered from cross-border environmental problems, owing to its geographic position, shape, and climatic conditions. The recent increase in the share

of Japanese ODA spent on environmental projects, especially in China, gives evidence of the increasing sensitivity in Japan to cross-border environmental problems, particularly those affecting Japan itself. Asuka-Zhang (1999) claims that in 1996 27% of total Japanese ODA was environmental ODA.

27. Rock (2000) mentions the significant role of influential overseas Chinese communities (particularly in the US, and at least since the mid 1970s) in putting pressure on the Taiwanese government to ecologize Taiwan's economy.

28. Apart from ASEAN and APEC, there is SAARC (the South Asian Association for Regional Cooperation), South Asia's counterpart of ASEAN. Its seven members agreed in 1995 to ratify the South Asian Preferential Trading Arrangement (SAPTA). In addition, several SREZs (Sub Regional Economic Zones), involving provinces or states of the nations in the region, are functioning. Their main focus is usually the transnational movement of capital, labor, and information, rather than trade in goods and services. Environmental considerations are rarely included.

29. This regional identity is only slowly "materializing" in strong regional environmental NGOs or cooperation between NGOs from individual Southeast Asian and East Asian countries (e.g. the People's Plan for the Twenty-First Century). Mittelman (1999) argues that civil society organizations from this regions have stronger links with non-governmental and governmental organizations in the North than with organizations in the region.

30. The member countries in 1998 are, in alphabetical order, Australia, Brunei, Canada, Chili, China, Hong Kong, Indonesia, Japan, Korea (South), Malaysia, Mexico, New Zealand, Papua New Guinea, Peru, Philippines, Russia, Singapore, Taiwan, Thailand, the US, and Vietnam.

Chapter 7

1. The treaty is meant to be open to all OECD members and to accession by non-member countries willing and able to meet the obligations. Nevertheless, the negotiations have been conducted exclusively among OECD members, with the justification that they account for the majority of world FDI outflows and inflows.

2. Ministerial Statement on the Multilateral Agreement on Investment (MAI), Paris, April 27–28, 1998. Environmental NGOs later claimed that consultations with interested parties in society and assessments of the consequences of MAI for the environment, for labor conditions, and for developing countries had taken place in only a limited number of countries.

3. For a critical environmental assessment of MAI, see the well-documented publications on the Web by the Environmental Defender's Office (Inquiry into the Multilateral Agreement on Investment, Sydney 1998; www.internetnorth.com.au/edo/edonsw/policy/mai.htm), by the Corporate Europe Observatory (MAI, The Reality of Six Months of "Consultation and Assessment," Amsterdam, 1998; www.interlog.com/~3mowchuck/news/mai), and by the Globalization and the MAI Information Centre (Writing the Constitution of a Single Global Economy: A concise Guide to the Multilateral Agreement on Investment—Supporters' and Opponents' Views, 1998; www.etk.net/preamble/maioverv.html).

4. This is the origin of the June 2000 demonstrations of French farmers and others against globalization and its core institutions (WTO, IMF, World Bank, and transnational companies as McDonalds) in a French provincial city Millau ("Seattle-sur-Tarn"). Following the WTO ruling, the US unilaterally announced trade sanctions against, among others, French local quality products (Roquefort cheese, pâté de foie gras, etc.). Protests of French farmers headed by the charismatic farmer José Bové resulted in attacks on one of the symbols of globalization and American domination: McDonalds fast-food restaurants. The legal trial against Bové and nine of his supporters in Millau brought over 30,000 protesters to Millau.

5. For updated information, see their Web site: http://www.web.net/cela/sbnn.htm.

6. The classical example is the "tuna-dolphin" case (GATT Docs. DS21/R, 3 September 1991 and DS29/R, 16 June 1994). Esty (1999a, p. 201) lists several WTO rulings in the tradition of the "tuna-dolphin" case, which aimed to prevent countries from limiting imports of goods produced in an "unnecessarily" environmentally harmful way (e.g., fur, shrimp, petroleum). This would, for instance, have major implications for eco-labeling schemes once they move from voluntary to more obligatory schemes, since these labels often include process and production methods standards in addition to product characteristics (Commission on Trade and Environment, WTO document press/TE 014, 14 November 1996). The WTO debate on eco-labeling has shown a clear North ("necessary") versus South ("discriminatory") division (Appleton 1999; Motaal 1999).

7. But this is beginning to change. The Biosafety Protocol, dealing with food products based on genetically modified organisms, was concluded in the framework of the UN Biodiversity Convention in late January 2000, which explicitly states that WTO agreements cannot overrule the Biosafety Protocol. The Protocol gives countries the right to refuse imports of biotechnologically produced products, using the precautionary principle, and demands the labeling of products based on genetically modified organisms within 2 years.

8. According to Brand (1999, pp. 286–287), this happened in the Basle Convention (banning trade between Annex VII and non-Annex VII countries), the International Convention for the Conservation of Atlantic Tuna (control of driftnet fishing), and also during the 1997 negotiation on the Kyoto Protocol, the discussions in 1998 on the Rotterdam Convention and the 1999 negotiations toward the Biosafety Protocol.

9. This committee, dominated by trade experts, had little if any results to show after having functioned for 5 years, Esty reports (1999a, p. 200).

10. Compare the positions of India, the US, and the EU regarding proposals to change article XX of GATT to prevent measures taken in the framework of multilateral environmental agreements from simply being blocked by referring to GATT (ADB 1997, p. 202).

11. Williams and Ford (1999) noted that the strategies of environmental movements toward the WTO have a similar duality. Part of the movement engages critically in reforming the WTO in more environmentally sound directions, while others reject the WTO and look for alternative institutional structures of global governance that give more precedence to environmental interests. Neither of the two strategies, the authors claimed, had had more than limited influence on actual developments.

12. Though Esty, one of the leading authors in this area, originally (1994) thought in the direction of a Global Environmental Organization to articulate environmental interests with respect to the WTO, he has more recently been arguing for a restructuring of the WTO as the most strategic and feasible option (Esty 1999a, p. 202ff.). A the same time, it should be noted that Chancellor Helmut Kohl proposed a World Environment Organization at the UN General Assembly Special Session, "Earth Summit 2" (June 1997), partly to balance the WTO.

13. Of the 141 developing countries, only 50 are full members of ISO. Only 25 participate in Technical Committee 207 (and its various subcommittees), which deals with most environmental standards. (Only 5 or 6 played an active role in the committee until 1995, and a few more in 1996 and 1997.) All developed countries are members, and almost all were actively involved in TC 207 negotiations from its establishment in 1993 on (Krut and Gleckman 1998, pp. 40–62).

14. Usually the following industrial sectors are included in this category of dirty industries: chemicals, pulp and paper, non-ferrous metals, iron and steel, non-metallic mineral products, and the oil and gas sector (Low and Yeats 1992; Jänicke 1995; Mani and Wheeler 1997).

15. Castleman's original report (1978) was followed by numerous other examples of individual polluting enterprises that moved to less developed countries. Leonard (1988, p. 68) lists some of these studies.

16. The decision on a major investment by the American chemical company Eastman Kodak in the Dutch Rijnmond industrial area in 1994 can be taken as an example. Eastman Kodak thought most investment factors at that location were favorable but initially considered the required environmental investments to be an obstacle to coming to the Netherlands. After the government of the Netherlands promised the company additional financial compensation (in the form of a favorable tax regime), Eastman Kodak decided to opt for this location despite the stringent environmental regime.

17. At first sight, this conclusion seems at odds with those presented by Low and Yeats and by Hettige, Lucas, and Wheeler. But differences in the aggregation levels, the reference years or period, the number and kind of countries involved, and the kind of products focused on make these studies hard to compare.

18. For empirical studies, see the references quoted in these publications.

19. The OECD (1997b) distinguishes among Market-seeking foreign direct investments (seeking opportunities to sell in oversees markets), which are not particularly sensitive to increased environmental costs; Production-platform-seeking FDI (serving regional markets by providing a platform for production and sales), which are also relatively insensitive to increased environmental costs; and Resource-seeking FDI (obtaining access to critical resources not available in their home markets), which are particularly susceptible to differences in environmental costs. The treaty is meant to be open to all OECD members and to accession by non-member countries willing and able to meet the obligations. Nevertheless, the negotiations have been conducted exclusively among OECD members, with the justification that they account for the majority of world FDI outflows and inflows.

Chapter 8

1. This chapter is based in part on earlier work (Mol and Frijns 1998; Frijns, Phung Thuy Phuong, and Mol 2000; Mol and van Vliet 1997a; Frijns et al. 1997). Numerous recent M.Sc. and Ph.D. thesis studies carried out at the environmental policy sub-department at Wageningen University have provided significant information and insights. See Wanyonyi, Onyango, and Wasonga on Kenya; Van Hengel, Le Van Khoa and Boot, Nguyen Phuc Quoc, Pham Thi Anh, Phung Thuy Phuong and Tran Thi My Dieu on Vietnam; Zhang Lei, Wang Xiaowen and Li Jungbing on China; van Vliet, Nieuwenhuis, Abeelen, and Kilsdonk and Le Jeune on Curaçao.

2. Only the four dragons, Taiwan, Korea, Singapore, and Hong Kong, are generally regarded as first-generation industrializing countries in East and Southeast Asia. The second tier contains countries as diverse as Indonesia, Malaysia, Thailand, China, and, I would argue, Vietnam (Hobday 1995).

3. Evans (1995) distinguishes three types of states in developing countries: developmental states, predatory states, and intermediate states. Predatory states are understood as states that extract resources at the expense of society, undercutting development even in the narrow sense of capital accumulation. Developmental states—of which the Vietnamese state can be seen as a representative—share high political capacity and relative autonomy, with close embeddedness in societal networks to effectively combine economic growth with political stability.

4. It is reported that in the Ho Chi Minh City region industrial growth in the first 10 months of 1998 (when the Asian financial crisis was in full swing) was still 12%. Before the Asian crisis, however, industrial growth figures for this region were far beyond the average for Vietnam (Mol and Frijns 1998).

5. Reviews on investments in 1997 show that Ho Chi Minh City harbored the greatest share of investments in Vietnam, with 591 projects and total invested capital of US $9,084,000,000; the capital, Hanoi, was second, with 289 projects and total capital of US $7,261,000,000; Dong Nai Province (near Ho Chi Minh City) was third, with 196 projects and total invested capital of US $4,093,000,000. Up to 96% of the capital invested in Dong Nai Province was supplied by foreign investors. Industry accounted for 61.9% of the total number of projects and 46.2% of the invested capital (Vietnam Investment Review, 17-5-1998). In Ho Chi Minh City, 40% of industrial investment was foreign. The Asian Development Bank (1998) reported similar increases but different absolute figures.

6. O'Rourke (2000) describes, among other examples, the successful community complaints at the Nike production site in Vietnam, which were backed by international and especially US protests against the poor working conditions for workers at Nike plants in developing countries.

7. Some national environmental NGOs have recently been established in China. Among these are the Center for Legal Assistance to Pollution Victims, the Beijing Foundation for Environmental Protection, the Research Center for Environment and Development, and the China Foundation for the Annual Evaluation of the Environment.

8. Like Vietnam, China shows substantial regional differences. Guangdong province in the south takes the lead, with foreign TNC production contributing to at least a third of total output (Held et al. 1999, p. 253).

9. Between 1992 and 1997, according to World Bank figures, approximately 5% of all World Bank loans to China were directed toward environmental protection (Jahiel 1998, p. 786). For a review of the considerable international efforts to stimulate cleaner production in Chinese industry from 1992 on, see Wang 1999.

10. According to Hein (1990b) and others, most micro-states are islands, and a considerable number of them can be characterized as development states or territories. Their special problems and disadvantages were put on the international agenda after a comprehensive document prepared by UNCTAD (1988) was discussed on Malta in 1988 and after the UN passed a resolution.

11. For various reasons, some internal and some external, Shell sold the refinery for the symbolic amount of one Dutch guilder (half a US dollar) to the government of the Netherlands Antilles and Curaçao, on the condition that Shell could not be held liable for any of the consequences of 70 years of oil refining on Curaçao, including the environmental consequences.

12. Since this is a small island with hardly any fresh water, large amounts of energy are used to desalinate sea water into fresh water for human consumption. This creates close relations between the water and energy networks (Nieuwenhuis 1997).

13. Of course, this growing international tourism has a number of other side effects, including effects on local culture and social cohesion.

14. The booming development of tourism on Aruba is often believed to be directly connected with the closure of the Esso refinery in the early 1980s. On Curaçao, too, some managers of major hotels located downwind from the Isla refinery question the feasibility of combining a major oil sector with international tourism on such a small island.

15. Hence, Curaçao government officials have shown little resistance against this move, although some policy makers and NGOs question whether this devolution of government control in tourism policy might prove to be an obstacle to a further move toward environmentally sound tourism development in the near future. The Minister of Economic Affairs is still formally the chairman of the CTDB, but his power is limited.

16. Lall (1992) argues that Côte d'Ivoire, Gabon, Mauritius, Zimbabwe, and Kenya are among the leading industrial countries in the 1980s, although the World Development Report (1997) statistics present a different picture based on value added by industry to GNP.

17. Whereas in the 1980s the structural adjustment policies were widely debated for their social effects, the emphasis in the debate shifted to the environmental consequences of these policies promoted by the World Bank and IMF during the 1990s. Despite sometimes heated debate on various economic and other results of structural adjustment policies (World Bank 1994; Castells 1996, pp. 148–150; Redclift and Sage 1998), a consensus seems to be emerging that structural adjustment policies can be both beneficial and destructive for the environment, depending on the local circumstances and the design of structural adjustment programs (Glover 1995).

18. With respect to air pollution, strict requirements or national standards were nonexistent up until 1997. Enforcement of the provisions was mainly conducted in response to public complaints, although inspectors and municipal authorities were allowed to carry out inspections and monitoring of air emissions (Frijns et al. 1997).

19. Although the major domestic groups (government officials, industries, NGOs, and experts) are quite well aware of the climate change issue—due to among other things the presence of UNEP and international donors—Kenya has not been very active in the field of domestic climate change policies. Kenya has limited influence in international negotiations due to lack of resources and because of low priority given to it. Effective coalitions within the G77 and the Organization of African Unity are seldom formed, partly because of competition for bilateral aid (Gupta 1997).

20. Early in 2000, Africa had only 2.8 million Internet connections; Europe (including Central and Eastern Europe but excluding CIS republics) had 88.2 million, the US and Canada 147.4 million, and Asia (excluding the Middle East) 65 million.

21. In 1994, around 9.7% of Kenyan GNP was Official Development Assistance (World Development report 1997). This does not put Kenya among the top African countries in terms of ODA, but it still means an increase since 1980 (when it was 5.6%). Since 1993, an overall global decline in ODA of 25% is paralleled by a sharp increase in the global flows of private investments. Global private investment flows to developing countries concentrated in three regions: East and Southeast Asia, Latin America and Central Europe (chapter 2 above; Gentry 1999).

22. A number of new concepts have been suggested to describe these processes, including negative sovereignty (Clapham 1996), the shadow state (Reno 1995), quasi-states (Jackson 1990), the vampire state (Frimpong-Ansah 1991), and the predatory state (Evans 1995).

Chapter 9

1. In the framework of the United Nation's International Human Dimensions Program, for instance, international relations scholars have more recently set up major networks for analyzing the effectiveness of the various multilateral environmental agreements, with the clear aim of enhancing the effectiveness of these agreements (Underdahl and Young 1996; Underdahl, Hisschemöller and von Moltke 1999; Young 1999).

2. An illustrative example might be the breakdown of the Global Climate Coalition, a group of large energy-intensive multinationals that claims that there is not enough scientific evidence on climate change to justify political measures such as those negotiated in Kyoto. After BP, Shell, and Dow Chemical, Ford left the coalition in December 1999, emphasizing the need for proactive measures and R&D.

3. The boundaries between these three principal domains—the economic, the political, and the socio-cultural—are of course also a matter of theoretical perspectives and analysis. Consumer demand, e.g., can be regarded as merely a market factor triggering reforms in various parts of the globe. But one can also say that growing environmental consciousness within civil society is at the origin of changing life styles, consumption patterns, and thus consumer demand.

4. To give one more example: the transnational food company Unilever decided in May 2000 to ban all genetically modified ingredients from their products sold in Europe. Unilever had been under consumer pressure in Germany, Austria, and the UK. As Unilever buys bulk ingredients such as soy centrally in Europe, it has one European strategy toward product-related environmental issues. Consumers of Unilever products in the Netherlands and Spain, for instance, "profit" from this economic mechanism, in contrast to consumers in the US.

5. In early 2000, a number of fast-food chains in the US, including McDonald's Corporation, ordered their French-fry suppliers to stop using genetically modified potatoes, such as those of Monsanto (*Wall Street Journal*, 28 April 2000). The US Department of Agriculture communicated on 31 March 2000 that all biotech corn varieties will drop from 33% in 1999 to 25% in 2000; all biotech cotton varieties will drop from 58% in 1999 to 45% in 2000; and all herbicide tolerant soy will drop from 57% in 1999 to 52% in 2000.

6. *De Volkskrant*, 19 May 2000, p. 13.

7. But, of course also resulting in major efforts by TNCs to legitimize their controversial products in different ways. Early in 2000, Monsanto, DuPont, Dow, Novartis, Zeneca, BASF, Aventis, and other corporations launched the Council for Biotechnology Information, which will spend US $50 million annually over a period of 5 years to win public acceptance of genetically engineered foods under the slogan "Good ideas are growing."

8. This issue is of course not restricted to global environmental governance, since it has to do with all kinds of legitimization issues connected to globalization patterns. It is not only a reformulation of Habermas's (1973) legitimization thesis on a global level but also a radicalization of it.

9. For the sake of the argument I will now concentrate on environmental reform processes and put between brackets those situations and processes marked by continuing environmental deterioration and failing reforms, such as in Kenya.

10. There is considerable debate on the extent to which these systems of innovations are indeed national, since, on the one hand, sectoral characteristics are often as important and, on the other, globalization patterns diminish the national specifics of these innovation systems. Nelson and Rosenberg (1993) make a threefold division between large high-income countries, small high-income countries and low-income countries. At the same time—and of relevance for environmental sociologists—the system of innovation relies heavily on the natural resource situation.

11. For more information on the struggle against Terminator, see the Web site of the Rural Advancement Foundation International: http://www.rafi.org11.

References

Abeelen, C. 1997. "Milieu onder het zand. Toerisme en duurzame ontwikkeling op Curaçao." In *Tussen Zandstrand en Asfaltmeer*, ed. A. Mol and B. van Vliet. Jan van Arkel.

A Blueprint for Survival. 1972. Penguin.

Adams, J. 1999. "Foreign direct investment and the environment: the role of voluntary corporate environmental management." Paper prepared for OECD Conference on Foreign Direct Investment and Environment, The Hague.

ADB (Asian Development Bank). 1996. *Asian Development Outlook 1996 and 1997*. Oxford University Press.

ADB. 1997. *Asian Development Outlook 1997 and 1998*. Oxford University Press.

ADB. 1998. *Asian Development Outlook 1998*. Oxford University Press.

Adeola, F. O. 2000. "Cross-national environmental injustice and human rights issues: A review of evidence in the developing world." *American Behavioral Scientist* 43, no. 4: 686–706.

Afash, S. 1998. Impact of Financial Crisis on Industrial Growth and Environmental Performance in Indonesia. US-Asia Environmental Partnership.

Afash, S., and D. Wheeler. 1998. Financial Crisis and Environmental Performance: Theoretical and Empirical Analysis. World Bank.

Agarwal, A., and S. Narain. 1991. Global Warming in an Unequal World: A Case of Environmental Colonialism. Centre for Science and Environment, New Delhi.

Albrow, M. 1997. "Travelling beyond local cultures: Socioscapes in a global city." In *Living the Global City*, ed. J. Eade. Routledge.

Amin, A. 1997. "Placing globalisation." *Theory, Culture and Society* 14, no. 2: 123–137.

Amin, A., and N. Thrift. 1994. "Living in the global." In *Globalisation, Institutions, and Regional Development in Europe*, ed. A. Amin and N. Thrift. Oxford University Press.

Andersen, M. S. 1994. *Governance by Green Taxes*: Manchester University Press.

Andersen, M. S., and J. D. Liefferink, eds. 1997. *European Environmental Policy: The Pioneers*: Manchester University Press.

Anderson, K., and D. H. Brooks. 1996. "Economic integration, cooperation and the Asian environment." *Asian Development Review* 14, no. 1: 44–71.

Angel, D., T. Feridhanusetyawan, and M. Rock. 1998. Towards Clean Shared Growth in Asia: A Policy and Research Agenda. Manuscript, Clark University, Worcester, Massachusetts.

Anton, D. J. 1995. Diversity, Globalisation and the Ways of Nature. International Development Research Centre.

Appleton, A. E. 1999. "Environmental labelling schemes: WTO law and developing country implications." In *Trade, Environment, and the Millennium*, ed. G. Sampson and W. Chambers. United Nations University Press.

Archer, M. 1991. "Sociology for one world: Unity and diversity." *International Sociology* 6, no. 2: 131–147.

Ariff, M. 1996. "Outlooks for ASEAN and NAFTA externalities." In *Cooperation or Rivalry?* ed. S. Nishijima and P. Smith. Westview.

Arrow, K., et al. 1995. "Economic growth, carrying capacity, and the environment." *Science* 268: 520–521.

Arts, B. 1998. *The Political Influence of Global NGOs: Case Studies on the Climate and Biodiversity Conventions.* Jan van Arkel/International Books.

Asuka-Zhang, S. 1999. "Transfer of environmentally sound technologies from Japan to China." *Environmental Impact Assessment Review* 19, no. 5/6: 553–567.

Audley, J. 1997. *Green Politics and Global Trade: NAFTA and the Future of Environmental Politics.* Georgetown University Press.

Axelrod, R. S. 1997. "Environmental policy and management in the European Union." In *Environmental Policy in the 1990s*, ed. N. Vig and M. Kraft. Congressional Quarterly Press.

Ayres, R. U., and U. E. Simonis, eds. 1995. Industrial Metabolism: Restructuring for Sustainable Development. United Nations University.

Bahro, R. 1977. *Die Alternative: Zur Kritik des real existierende Szialismus.* Europaïsche Verlaganstalt.

Bahro, R. 1984. *From Red to Green.* Verso.

Barber, B. R. 1995. *Jihad vs. McWorld.* Random House.

Bauman, Z. 1993. *Postmodern Ethics.* Blackwell.

Baumol, W., and W. Oates. 1988. *The Theory of Environmental Policy.* Cambridge University Press.

Beck, U. 1986. *Risikogesellschaft: Auf dem Weg in eine andere Moderne.* Suhrkamp.

Beck, U. 1991. *Politik in der Risikogesellschaft.* Suhrkamp.

Beck, U. 1994a. "The reinvention of politics: Towards a theory of reflexive modernisation." In *Reflexive Modernisation*, ed. U. Beck et al. Polity Press.

Beck, U. 1994b. "Replies and critiques. Self-dissolution and self-endangerment of industrial society: What does this mean?" In *Reflexive Modernisation*, ed. U. Beck et al. Polity Press.

Beck, U. 1996. "World risk society as cosmopolitan society? Ecological questions in a framework of manufactured uncertainties." *Theory, Culture and Society* 13, no. 4: 1–32.

Beck, U. 1997. *Was ist Globalisierung? Irrtümer des Globalismus—Antworten auf Globalisierung*. Suhrkamp.

Bell, D. 1976. *The Coming of Post-Industrial Society: A Venture in Social Forecasting*. Basic Books.

Bennell, P. 1995. "British manufacturing investment in sub-Saharan Africa: Corporate responses during structural adjustment." *Journal of Development Studies* 31, no. 2: 195–217.

Blühdorn, I. 2000. "Ecological modernisation and post-ecologist politics." In *Environment and Global Modernity*, ed. G. Spaargaren et al. Sage.

Boehmer-Christiansen, S., and J. Murphy. 1997. "Ecological modernisation—whose capacity is being built? Conflicting evidence from the UK." In *Umweltpolitik und Staatsversagen*, ed. L. Metz and H. Weidner. Sigma.

Boli, J., and G. M. Thomas. 1997. "World culture in the world polity: A century of international non-governmental organization." *American Sociological Review* 62, April: 171–190.

Bonnano, A., et al., eds. 1994. *From Columbus to ConAgra: The Globalisation of Agriculture and Food*. University Press of Kansas.

Bookchin, M. 1982. *The Ecology of Freedom: The Emergence and Dissolution of Hierarchy*. Cheshire Books.

Boons, F. 1997. "Organisatieverandering en ecologische modernisering: het voorbeeld van groene productontwikkeling." Paper prepared for Dutch NSAV conference.

Bosso, C. J. 1988. "Transforming adversaries into collaborators: Interest groups and the regulation of chemical pesticides." *Policy Sciences* 21: 3–22.

Brack, D. 1999. "Environmental treaties and trade: Multilateral environmental agreements and the multinational trading system." In *Trade, Environment, and the Millennium*, ed. G. Sampson and W. Chambers. United Nations Press.

Braudel, F. 1979. *Civilisation matérielle, économique et capitalisme, XVe-XVIIIe siècle*. Colin.

Brautigam, D. 1994. "African industrialization in comparative perspective: The question of scale." In *African Capitalists in African Development*, ed. B. Berman and C. Leys. Lynne Rienner.

Bray, M., and S. Packer. 1993. *Education in Small States*. Pergamon.

Briguglio, L. 1995. "Small island developing states and their economic vulnerabilities." *World Development* 23, no. 9: 1615–1632.

Broadbent, J. 1998. *Environmental Politics in Japan: Networks of Power and Protest*. Cambridge University Press.

Brown, H. S., J. J. Himmelberger, and A. L. White. 1993. "Development-environment interactions in the export of hazardous technologies." *Technological Forecasting and Social Change* 42: 125–155.

Brown, L. R. 1972. *World without Borders*. Random House.

Bunker, S. G. 1996. "Raw material and the global economy: Oversights and distortions in industrial ecology." *Society and Natural Resources* 9: 419–429.

Burnett, A. 1992. *The Western Pacific: Challenge of Sustainable Growth*. Earthscan.

Buttel, F. 1992. "Environmentalization: Origins, processes and implications for rural social change." *Rural Sociology* 57, no. 1: 1–27.

Buttel, F. 2000. "Classical theory and contemporary environmental sociology." In *Environment and Global Modernity*, ed. G. Spaargaren et al. Sage.

Buttel, F. H., A. P. Hawkins, and A. G. Power. 1990). "From limits to growth to global change. Constraints and contradictions in the evolution of environmental science and ideology." *Global Environmental Change*, December 1990: 57–66.

Cameron, O. 1996. "Japan and South-East Asia's environment." In *Environmental Change in South-East Asia*, ed. M. Parnwell and R. Bryant. Routledge.

Cameron, J. 2000. "The precautionary principle." In *Trade, Environment, and the Millennium*, ed. G. Sampson and W. Chambers. United Nations University Press.

Cao Van Sung, ed. 1995. *Ressource Biologiques et Environnement au Vietnam: Réalité et Perspectives*. Hanoi: Gioi.

Castells, M. 1996. *The Rise of the Network Society*. Blackwell.

Castells, M. 1997a. *The Power of Identity*. Blackwell.

Castells, M. 1997b. *End of Millennium*. Blackwell.

Castleman, B. I. 1979a. "The export of hazardous factories to developing nations." *International Journal of Health Services* 9, no. 4: 569–606.

Castleman, B. I. 1979b. "Exporting hazardous industries." *Ecologist* 9, no. 3: 80–85.

Cavanagh, J., D. Wysham, and M. Arruda, eds. 1994. *Beyond Bretton Woods: Alternatives to the Global Economic Order*. Pluto.

Chandra, R. 1992. *Industrialisation and Development in the Third World*. Routledge.

Charnovitz, S. 1994. "NAFTA's social dimension: Lessons from the past and framework for the future." *International Trade Journal* 8, no. 39.

Christoff, P. 1996. "Ecological modernisation, ecological modernities." *Environmental Politics* 5, no. 3: 476–500.

Clapham, C. 1996. *Africa and the International System: The Politics of State Survival*. Cambridge University Press.

Clark, G. L. 1996. "Global competition and environmental regulation: Is the 'race to the bottom' inevitable?" In *Markets, the State and the Environment*, ed. R. Eckersley. Macmillan.

Clarke, G. 1998. *The Politics of NGOs in South-East Asia: Participation and Protest in the Philippines*. Routledge.

Cohen, M. A. 1996. "The hypothesis of urban convergence: Are cities in the North and South becoming more alike in an age of globalisation?" In *Preparing for the Urban Future*, ed. M. Cohen et al. Woodrow Wilson Center Press.

Cohen, M. J. 2000. "Ecological modernisation, environmental knowledge and national character: A preliminary analysis of the Netherlands." *Environmental Politics 9*, no. 1: 77–105.

Commission on Global Governance. 1995. *Our Global Neighbourhood*. Oxford University Press.

CTDB (Curaçao Tourism Development Board). 1993. A Master Plan for Tourism Development.

Dalby, S. 1996. "Crossing disciplinary boundaries: Political geography and international relations after the Cold War." In *Globalisation*, ed. E. Kofman and G. Youngs. Pinter.

Daly, H., and J. Cobb. 1994. *For the Common Good: Redirecting the Economy toward Community, the Environment and Sustainable Development*. Beacon.

Dasgupta, S., and D. Wheeler. 1996. Citizens Complaints as Environmental Indicators: Evidence from China. Policy Research Department, World Bank.

Dasgupta, S., A. Mody, S. Roy, and D. Wheeler. 1995. "Environmental regulation and development: A cross-country empirical analysis." Working paper 1448, Policy Research Department, World Bank.

Davey, W. J. 1999. "The WTO dispute settlement system." In *Trade, Environment, and the Millennium*, ed. G. Sampson and W. Chambers. United Nations University Press.

Desta, Z. 1996. Industrial Environmental Management: The Case of Awassa Textile Factory. M.Sc. thesis, Wageningen University.

Di Chang-Xing. 1999. "ISO 14001: The severe challenge for China." In *Growing Pains*, ed. W. Wehrmeyer and Y. Mulugetta. Greenleaf.

Dicken, P. 1997. "Transnational corporations and nation-states." *International Social Science Journal* 151 (March 1997. pp. 77–89.

Dicken, P. 1998. *Global Shift: Transforming the World Economy*. Third edition. Paul Chapman/Sage.

Dickens, P. 1998. "Beyond sociology: Marxism and the environment." In *The International Handbook of Environmental Sociology*, ed. M. Redclift and G. Woodgate. Elgar.

Dorfman, R. 1991. "Protecting the global environment: A modest proposal." *World Development* 19, no. 1: 103–110.

Doyle, E. 1999. An Evaluation of Stakeholder Participation in Private Sector Planning for an Environmentally Sensitive Energy Development. M.Sc. thesis, Centre for the Urban Environment, Rotterdam/Wageningen.

Dryzek, J. S. 1997. *The Politics of the Earth: Environmental Discourses*. Oxford University Press.

Duyvendak, W., I. Horstik, and B. Zagema, eds. 1999. Het Groene Poldermodel: Consensus en conflict in de milieupolitiek. Instituut voor Publiek en Politiek/Vereniging Milieudefensie, Amsterdam.

Eccleston, B., and D. Potter. 1996. "Environmental NGOs and different political contexts in South-East Asia: Malaysia, Indonesia and Vietnam." In *Environmental Change in South-East Asia*, ed. M. Parnwell and R. Bryant. Routledge.

Edquist, C., ed. 1997. *System of Innovation*. Pinter.

Ekins, P. 1995. "The Kuznets curve for the environment and economic growth: Examining the evidence." Unpublished paper, Birkbeck College, University of London.

Elliott, L. 1998. *The Global Politics of the Environment*. New York University Press.

Enzensberger, H. M. 1974. "A critique of political ecology." *New Left Review* 84: 3–32.

Esty, D. C. 1994. Greening the GATT: Trade, Environment, and the Future. Institute for International Economics.

Esty, D. C. 1999a. "Economic integration and the environment." In *The Global Environment*, ed. V. Kraft and R. Axelrod. Congressional Quarterly.

Esty, D. C. 1999b. "Environmental governance at the WTO: Outreach to civil society." In *Trade, Environment, and the Millennium*, ed. G. Sampson and W. Chambers. United Nations University Press.

European Environmental Agency. 1998. *Europe's Environment: The Second Assessment*. Elsevier.

Evans, P. 1995. *Embedded Autonomy: States and Industrial Transformation*. Princeton University Press.

Evans, P. 1996. "Government action, social capital and development: Reviewing the evidence on synergy." *World Development* 24, no. 6: 1119–1132.

Falke, J. 1996. "Integrating scientific expertise into regulatory decision-making: The role of non-governmental standardization organisations in the regulation of risks to health and the environment." Working Paper RSC 96/9, European University Institute, Florence.

Featherstone, M., ed. *Global Culture: Nationalism, Globalization and Modernity*. Sage.

Fisher-Kowalski, M. 1997. "Tons, joules, and money: Modes of production and their sustainability problems." *Society and Natural Resources* 10, no. 1: 61–85.

Frank, D. J., A. Hironaka, and E. Schofer. 2000. "The nation-state and the natural environment over the twentieth century." *American Sociological Review* 65, no. 1: 77–95.

Friends of the Earth Netherlands. 1992. Sustainable Netherlands.

Frijns, J., and J. Malombe, eds. 1997. *Cleaner Production and Small Scale Enterprise Development in Kenya*. CUE/HABRI (Rotterdam/Nairobi).

Frijns, J., et al. 1997. *Pollution Control of Small Scale Metal Industries in Nairobi*. WAU/HABRI (Wageningen/Nairobi).

Frijns, J., Phung Thuy Phuong, and A. P. J. Mol. 2000. "Ecological modernisation theory and industrialising economies: The case of Viet Nam." *Environmental Politics* 9, no. 1: 257–292.

Frimpong-Ansah, J. H. 1991. *The Vampire State in Africa: The Political Economy of Decline in Ghana*. James Curley.

Gentry, B. 1999. "Foreign direct investment and the environment: Boon or bane?" Paper prepared for OECD Conference on Foreign Direct Investment and Environment, The Hague.

Giddens, A. 1982. *Profiles and Critiques in Social Theory*. Macmillan.

Giddens, A. 1984. *The Constitution of Society*. Polity Press.

Giddens, A. 1990. *The Consequences of Modernity*. Polity Press.

Giddens, A. 1991. *Modernity and Self-Identity: Self and Society in the Late Modern Age*. Polity Press.

Giddens, A. 1994a. *Beyond Left and Right: The Future of Radical Politics*. Polity Press.

Giddens, A. 1994b. "Replies and critiques. Risk, trust, reflexivity." In *Reflexive Modernisation*, ed. U. Beck et al. Polity Press.

Gille, Z. 2000. "Legacy of waste or wasted legacy? The rise and fall of industrial ecology in socialist and postsocialist Hungary." *Environmental Politics* 9, no. 1: 203–231.

Glover, D. 1995. "Structural adjustment and the environment." *Journal of International Development* 7, no. 2: 285–289.

Glyn, A., and B. Sutcliffe. 1992. "'Global but leaderless'? The new capitalist order." In *New World Order: The Socialist Register*, ed. R. Miliband and L. Panitch. Merlin.

Goldblatt, D. 1996. *Social Theory and the Environment*. Polity Press.

Goldeman, G. 1999. "The environmental implications of foreign direct investment: Policy and institutional issues." Paper prepared for OECD Conference on Foreign Direct Investment and Environment, The Hague.

Goldfrank, W. L., D. Goodman, and A. Szasz, eds. 1999. *Ecology and the World-System*. Greenwood.

Gordon, D. M. 1988. "The global economy: New edifice or crumbling foundations?" *New Left Review* 168: 24–64.

Gorz, A. 1989. *Critique of Economic Reason*. Verso.

Gouldson, A., and J. Murphy. 1996. "Ecological modernization and the European Union." *Geoforum* 27, no. 1: 11–21.

Gourevitch, P. 1978. "The second image reversed: The international sources of domestic politics." *International Organization* 32, no. 4: 21–68.

Grant, W. 1993. "Transnational companies and environmental policy-making: The trend of globalisation." In *European Integration and Environmental Policy*, ed. J. Liefferink et al. Belhaven.

Griesgraber, M. J., and B. G. Gunter, eds. 1995. *Promoting Development: Effective Global Institutions for the 21st Century*. Pluto.

Griesgraber, M. J., and B. G. Gunter, eds. 1996. *Development: New Paradigms and Principles for the 21st Century*. Pluto.

Group of Lisbon. 1995. *Limits to Competition*. MIT Press.

Grubb, M., C. Vrolijk, and D. Brack. 1999. The Kyoto Protocol: A Guide and Assessment. Royal Institute of International Affairs.

Guha, R. 1989. "Radical American environmentalism and wilderness preservation: A Third World critique." *Environmental Ethics* 11: 71–83.

Gupta, J. 1997. The Climate Change Convention and Developing Countries: From Conflict to Consensus? Dissertation, Free University, Amsterdam.

Haas, P. M. 1990. *Saving the Mediterranean: The Politics of International Environmental Cooperation.* Columbia University Press.

Haas, P. M., R. O. Keohane, and M. A. Levy. 1993. *Institutions for the Earth: Sources of Effective Environmental Protection.* MIT Press.

Habermas, J. 1973. *Legitimationsprobleme im Spätkapitalismus.* Suhrkamp.

Hajer, M. A. 1995. *The Politics of Environmental Discourse: Ecological Modernisation and the Policy Process.* Clarendon.

Hajer, M. A. 1996. "Ecological modernization as cultural politics." In *Risk, Environment and Modernity,* ed. S. Lash et al. Sage.

Hannigan, J. A. 1995. *Environmental Sociology: A Social Constructivist Perspective.* Routledge.

Harris, J. M. 1992. "Global institutions for sustainable development." In *Sustainability and Environmental Policy,* ed. F. Dietz et al. Sigma.

Harris, S. 1996. The Search for a Landfill Site in an Age of Risk: The Role of Trust, Risk and the Environment. Dissertation, McMaster University.

Harvey, D. 1989. *The Condition of Postmodernity.* Blackwell.

HDP (Human Dimensions Program). 1996. Industrial Transformation. Scoping report, Institute for Environmental Studies, Free University, Amsterdam.

Heerings, H. 1993. "The role of environmental policies in influencing patterns of investment of transnational corporations: Case study of the phosphate fertilizer industry." In Environmental Policies and Industrial Competitiveness. OECD.

Hein, P. L. 1990a. "Between Aldabra and Nauru." In *Sustainable Development and Environmental Management of Small Islands,* ed. W. Beller et al. UNESCO and Parthenon.

Hein, P. L. 1990b. "Economic problems and prospects of small islands." In *Sustainable Development and Environmental Management of Small Islands,* ed. W. Beller et al. UNESCO and Parthenon.

Held, D. 1995. *Democracy and the Global Order: From the Modern State to Cosmopolitan Governance.* Polity Press.

Held, D., A. McGrew, D. Goldblatt, and J. Perraton. 1999. *Global Transformations: Politics, Economics and Culture.* Stanford University Press.

Henson, S., R. Loader, A. Swinbank, and M. Bredahl. 1999. The Impact of Sanitary and Phytosanitary Measures on Developing Country Exports of Agricultural and Food Products. WTO/World Bank.

Hesselberg, J. 1995. "Transnational companies and industrial pollution in the South: An overview." In *The Greening of Industry Resource Guide and Bibliography*, ed. P. Groenewegen et al. Island.

Hettige, H., R. E. B. Lucas, and D. Wheeler. 1992. "The toxic intensity of industrial pollution: Global patterns, trends and trade policy." *American Economic Review* 82, no. 2: 478–481.

Hettige, H., M. Huq, S. Pargal, and D. Wheeler. 1996. "Determinants of pollution abatement in developing countries: Evidence from South and Southeast Asia." *World Development* 24, no. 12: 1891–1904.

Higgott, R. 1999. "The political economy of globalisation in East Asia. The salience of 'region building.'" In *Globalisation and the Asia-Pacific*, ed. K. Olds et al. Routledge.

Hirst, P., and G. Thompson. 1996. *Globalisation in Question?* Polity Press.

Hobday, M. 1995. "East Asian latecomer firms: Learning the technology of electronics." *World Development* 23, no. 7: 1171–1193.

Hobsbawm, E. 1994. *Age of Extremes: The Short Twentieth Century, 1914–1991.* Michael Joseph.

Ho Chi Minh Construction Department. 1996. Scheme of Planning Concentrated Industrial Zones in Ho Chi Minh City, volume 1-3a.

Hogenboom, B. B. 1998. *Mexico and the NAFTA Environment Debate: The Transnational Politics of Economic Integration.* Jan van Arkel/International Books.

Hogenboom, J., A. P. J. Mol, and G. Spaargaren. 2000. "Dealing with environmental risk in reflexive modernity." In *Risk in the Modern Age*, ed. M. Cohen. Macmillan.

Hoogvelt, A. 1997. *Globalisation and the Postcolonial World: The New Political Economy of Development.* Macmillan.

Hovden, E. 1999. "As if nature doesn't matter: Ecology, regime theory and international relations." *Environmental Politics* 8, no. 2: 50–74.

Huber, J. 1982. *Die verlorene Unschuld der Ökologie: Neue Technologien und superindustrielle Entwicklung.* Fisher.

Huber, J. 1985. *Die Regenbogengesellschaft: Ökologie und Sozialpolitik.* Fisher.

Huber, J. 1991. *Unternehmen Umwelt: Weichenstellungen für eine ökologische Marktwirtschaft.* Fisher.

Huber, J. 1993. "Ökologische Modernisierung. Bedingungen des Umwelthandelns in den neuen und alten Bundesländern." *Kölner Zeitschrift für Soziologie und Sozialpsychologie* 45, no. 2: 288–304.

Hudson, S. 1993. "Exploring the relationship between investment, trade and environment." In Environmental Policies and Industrial Competitiveness. OECD.

Huq, A., B. N. Lohani, K. F. Jalal, and E. A. R. Ouano. 1999. "The Asian Development Bank's role in promoting cleaner production in the People's Republic of China." *Environmental Impact Assessment Review* 19, no. 5/6: 541–552.

Hurley, A. 1995. *Environmental Inequalities: Class, Race and Industrial Pollution in Gary, Indiana, 1945–1980.* University of North Carolina Press.

Independent Commission on International Development Issues. 1980. North-South, a Programme for Survival: Report of the Independent Commission on International development Issues.

Inglehart, R. 1990. *Culture Shift in Advanced Industrial Society.* Princeton University Press.

IEA (International Energy Agency). 1998. CO_2 Emissions from Fuel Combustion 1971–1996.

Irvin, G. 1996. "Emerging issues in Viet Nam: Privatisation, equality and sustainable growth." *European Journal of Development Research* 8, no. 2: 178–199.

Jackson, R. H. 1990. *Quasi-States: Sovereignty, International Relations and the Third World.* Cambridge University Press.

Jacobson, H. K. 1984. *Networks of Interdependence: International Organisations and the Global Political System.* Second edition. Knopf.

Jahiel, A. R.,. 1998. "The organization of environmental protection in China." *China Quarterly*, December: 757–787.

Jänicke, M. 1978. "Blauer Himmel über den Industriestädten—ein optische Täuchung." In *Umweltpolitik*, ed. M. Jänicke. Opladen.

Jänicke, M. 1986. *Staatsversagen: Die Ohnmacht der Politik in die Industriegesellschaft.* Piper.

Jänicke, M. 1990. "Erfolgsbedingungen von Umweltpolitik im internationalen Vergleich." *Zeitschrift für Umweltpolitik und Umweltrecht* 3: 213–232.

Jänicke, M. 1991. The Political System's Capacity for Environmental Policy. Freie Universität Berlin.

Jänicke, M. 1993. "Über ökologische und politieke Modernisierungen." *Zeitschrift für Umweltpolitik und Umweltrecht* 2: 159–175.

Jänicke, M. 1997. "Umweltpolitik: Global am Ende oder am Ende Global?" In *Politik der Globalisierung*, ed. U. Beck. Suhrkamp.

Jänicke, M., et al. 1992. *Umweltentlastung durch industriellen Strukturwandel? Eine explorative Studie über 32 Industrieländer. 1970 bis 1990.* Sigma.

Jänicke, M., et al. 1995. "Green industrial policy and the future of 'dirty industries.'" Unpublished paper, Freie Universität Berlin.

Jansen, K. 1997. "Economic reform and welfare in Viet Nam." Working paper 260, Institute of Social Sciences,The Hague.

Jessop, B. 1984. *The Capitalist State: Marxist Theories and Methods.* Blackwell.

Johnson, C. 1982. *MITI and the Japanese Miracle: The Growth of Industrial Policy, 1925–1975.* Stanford University Press.

Johnson, P. M., and A. Beaulieu. 1996. *The Environment and NAFTA: Understanding and Implementing the New Continental Law.* Island.

Johnson, S. P., and G. Corcelle. 1989. *The Environmental Policy of the European Communities*. Graham and Trotman.

Johnston, A. I. 1988. "China and international environmental institutions: A decision rule analysis." In *Energizing China*, ed. M. McElroy et al. Harvard University Press.

Jokinen, P. 2000. Europeanisation and ecological modernisation: Agri-environmental policy and practices in Finland. *Environmental Politics* 9, no. 1: 138–170.

Jokinen, P., and K. Koskinen,. 1998. "Unity in environmental discourse? The role of decision-makers, experts and citizens in developing Finish environmental policy." *Policy and Politics* 26, no. 1: 55–70.

Jones, M. 1997. "The role of stakeholder participation: Linkages to stakeholder impact assessment and social capital in Camisea, Peru." *Greener Management International* 19: 87–97.

Karliner, J. 1997. *The Corporate Planet: Ecology and Politics in the Age of Globalization*. Sierra Club Books.

Kenya Government. 1996. "Industrial transformation to the year 2020." Sessional paper 2, Government of Kenya, Nairobi.

Keohane, R. O., and M. A. Levy. 1996. *Institutions for Environmental Aid: Pitfalls and Promise*. MIT Press.

Keohane, R. O., and J. S. Nye. 1977. *Power and Interdependence: World Politics in Transition*. Little, Brown.

Kolk, A. 1999. "Environmental management and organisational change: The impact of the World Bank." In *Growing Pains*, ed. W. Wehrmeyer and Y. Mulugetta. Greenleaf.

Korten, D. C. 1995. *When Corporations Rule the World*. Earthscan.

Krut, R., and H. Gleckman. 1998. *ISO 14001: A Missed Opportunity for Sustainable Global Industrial Development*. Earthscan.

Kumar, A., J. Milner, and A. Petsonk. 1996. "The North American Free Trade Association." In *Greening International Institutions*, ed. J. Werksman. Earthscan.

Kumar, K. 1995. *From Post-Industrial to Post-Modern Society: New Theories of the Contemporary World*. Blackwell.

Lall, S. 1992. "Structural Problems of African Industry." In *Alternative Development Strategies in Sub-Saharan Africa*, ed. F. Stewart et al. Macmillan.

Lash, S., and J. Urry. 1994. *Economies of Signs and Space*. Sage.

La Spina, A., and G. Scortino. 1993. "Common agenda, Southern rules: European integration and environmental change in the Mediterranean states." In *European Integration and Environmental Policy*, ed. J. Liefferink et al. Belhaven.

Leff, E. 1995. *Green Production: Toward an Environmental Rationality*. Guilford.

Leonard, H. J. 1988. *Pollution and the Struggle for the World Product: Multinational Corporations, Environment and International Comparative Advantage*. Cambridge University Press.

LeQuesne, C. 1996. Reforming World Trade: The Social and Environmental Priorities. Oxfam.

Leroy, P., and J. van Tatenhove. 2000. "New policy arrangements in environmental politics: The relevance of political and ecological modernization." In *Environment and Global Modernity*, ed. G. Spaargaren et al. Sage.

Leventein, C., and S. W. Eller. 1980. "Are hazardous industries fleeing abroad?" *Business and Society Review* 34: 44–46.

Levy, M. A., and O. R. Young, eds. 1996. *The Effectiveness of International Regimes*. Cornell University Press.

Liefferink, J. D. 1997. *Environment and the Nation-State: The Netherlands, the European Union and Acid Rain*. Manchester University Press.

Liefferink, J. D., P. D. Lowe, and A. P. J. Mol, eds. 1993. *European Integration and Environmental Policy*. Belhaven.

Lindblom, C. E. 1959. "The science of 'muddling through.'" *Public Administration Review* 19, no. 9: 79–88.

Lindhqvist, T., and R. Lifset. 1998. "Getting the goal right: EPR and DfE." *Journal of Industrial Ecology* 2, no. 1: 6–8.

Lipietz, A. 1987. *Mirages and Miracles: The Crisis of Global Fordism*. Verso.

Lipschutz, R. 1996. *Global Civil Society and Global Environmental Governance*. State University of New York Press.

Lloyd, P. J. 1994. "Intraregional trade in the Asian and Pacific region." *Asian Development Review* 12, no. 2: 113–143.

Low, P., ed. 1992. "International trade and the environment." Discussion paper 159, World Bank.

Low, P., and A. Yeats. 1992. "Do 'Dirty' Industries Migrate?" In "International trade and the environment," World Bank discussion paper 159, ed. P. Low.

Lubeck, P. M., and M. J. Watts. 1994. "An alliance of oil and maize? The response of indigenous and state capital to structural adjustment in Nigeria." In *African Capitalists in African Development*, ed. B. Berman and C. Leys. Lynne Rienner.

Lucas, R. E. B., D. Wheeler, and H. Hettige. 1992. "Economic development, environmental regulation and the international migration of toxic industrial pollution: 1960–1988." In "International trade and the environment," World Bank discussion paper 159, ed. P. Low.

Luhmann, N. 1986. *Ökologische Kommunikation*. Westdeutscher Verlag.

Lundqvist, L 2000. "Capacity-building or social construction? Explaining Sweden's shift towards ecological modernisation." *Geoforum* 31, no. 1: 21–32.

Maddock, R. T. 1995. "Environmental security in East Asia." *Contemporary Southeast Asia* 17, no. 1: 20–37.

Mani, M., and D. Wheeler. 1997. "In search of pollution havens? Dirty industry in the world economy, 1960–1995." Working paper, Policy Research Department, World Bank.

Marsden, T., et al. 1993. *Constructing the Countryside*. University College London Press.

Martin, H.-P., and H. Schumann. 1996. *Die Globalisierungsfalle: Der Angriff auf Demokratie und Wohlstand*. Rowohlt.

Maull, H. W. 1992. "Japan's global environmental policies." In *The International Politics of the Environment*, ed. A. Hurrell and B. Kingsbury. Clarendon.

Maxeiner, D., and M. Miersch. 1996. *Öko-Optimismus: Leben Im 21sten Jahrhundert*. Metropolitan Verlag.

McAfee, K. 1999. Biodiversity and the Contradictions of Green Developmentalism. Dissertation, University of California, Berkeley.

McCormick, J. 1999. "The role of environmental NGOs in international regimes." In *The Global Environment*, ed. N. Vig and R. Axelrod. Congressional Quarterly Press.

McDowell, M. A. 1990. "The development of the environment in ASEAN." *Pacific Affairs* 62, no. 3: 307–329.

McElroy, J. L., B. Potter, and E. Towle. 1990. "Challenges for sustainable development in small caribbean islands." In *Sustainable Development and Environmental Management of Small Islands*, ed. W. Beller et al. UNESCO and Parthenon.

McElroy, M. B., C. P. Nielsen, and P. Lydon, eds. 1998. *Energizing China: Reconciling Environmental Protection and Economic Growth*. Harvard University Press.

McGrew, A. 1992. "A global society?" In *Modernity and Its Futures*, ed. S. Hall et al. Polity Press.

McMichael, P. 1996a. "Globalisation: Myths and realities." *Rural Sociology* 61, no. 1: 25–55.

McMichael, P. 1996b. *Development and Social Change: A Global Perspective*. Pine Forge Press.

Meyer, J. W. 1987. "The world polity and the authority of the nation-state." In *Institutional Structure*, ed. G. Thomas et al. Sage.

Miller, M. A. L. 1995. *The Third World in Global Environmental Politics*. Open University Press.

Mitsuda, H. 1997. "Surging environmentalism in Japan: A sociological perspective." In *International Handbook of Environmental Sociology*, ed. M. Redclift and G. Woodgate. Elgar.

Mittelman, J. H. 1998. "Globalisation and environmental resistance politics." *Third World Quarterly* 19, no. 5: 847–872.

Mittelman, J. H. 1999. "Resisting globalisation: Environmental politics in Eastern Asia." In *Globalisation and the Asia-Pacific*, ed. K. Olds et al. Routledge.

Mlinar, Z. 1992. "Individuation and globalisation: The transformation of territorial social organization." In *Globalisation and Territorial Identities*, ed. Z. Mlinar. Avebury.

Mol, A. P. J. 1995. *The Refinement of Production: Ecological Modernisation Theory and the Chemical Industry.* Jan van Arkel/International Books.

Mol, A. P. J. 1996. "Ecological modernisation and institutional reflexivity. Environmental reform in the late modern age." *Environmental Politics* 5, no. 2: 302–323.

Mol, A. P. J. 1999a. "A Globalização e a mundança dos modelos de controle e poluição industrial: a teoria da modernização ecológica." In *Qualidade de vida & riscos ambientais*, ed. S. Herculano et al. Editoria da Universidade Federal Fluminense.

Mol, A. P. J. 1999b. "Ecological modernisation and the environmental transition of Europe: Between national variations and common denominators." *Journal of Environmental Policy and Planning* 1, no. 2: 17–38.

Mol, A. P. J. 2000. "The environmental movement in an era of ecological modernisation." *Geoforum* 31: 45–56.

Mol, A. P. J., and J. Frijns. 1998. "Environmental reforms in industrial Vietnam: The Ho Chi Minh City region." *Asia-Pacific Development Journal* 5, no. 2: 117–138.

Mol, A. P. J., and J. D. Liefferink. 1993. "European environmental policy and global interdependence: A review of theoretical approaches." In *European Integration and Environmental Policy*, ed. J. Liefferink et al. Belhaven.

Mol, A. P. J., and D. S. Sonnenfeld, eds. 2000. *Ecological Modernisation Around the World: Perspectives and Critical Debates.* Frank Cass.

Mol, A. P. J., and G. Spaargaren. 1993. "Environment, modernity and the risk-society: The apocalyptic horizon of environmental reform." *International Sociology* 8, no. 4: 431–459.

Mol, A. P. J., and G. Spaargaren. 2000. "Ecological modernization theory in debate: A review." *Environmental Politics* 9, no. 1: 17–49.

Mol, A. P. J., and G. Spaargaren. 2001. "The environmental state in transition. Exploring the contradictions and commonalities between ToP and EMT." *Organization and Environment* 14, no. 3.

Mol, A. P. J., and B. van Vliet. 1997a. *Tussen Zandstrand en Asfaltmeer: Milieubeheer op Curaçao.* Jan van Arkel.

Mol, A. P. J., and B. van Vliet. 1997b. "Epiloog. Naar een duurzaam Curaçao." In *Tussen Zandstrand en Asfaltmeer*, ed. A. Mol and B. van Vliet. Jan van Arkel.

Mol, A. P. J., et al. 1996. "Joint environmental policy-making in comparative perspective." Paper presented at Greening of Industry conference, Heidelberg.

Mol, A. P. J., V. Lauber, and J. D. Liefferink, eds. 2000. *The Voluntary Approach to Environmental Policy.* Oxford University Press.

Morales, I. 1999. "NAFTA: The governance of economic openness." *Annals of the American Academy of Political Social Scientists* 565, September: 35–65.

Morse, E. L. 1976. *Modernization and the Transformation of International Relations.* Free Press.

Moser, T., and D. Miller. 1997. "Multinational corporations' impacts on the environment and communities in the developing world: A synthesis of the contemporary debate." *Greener Management International* 19: 40–51.

Motaal, D. A. 1999. "The agreement on technical barriers to trade, the committee on trade and environment, and eco-labelling." In *Trade, Environment, and the Millennium*, ed. G. Sampson and W. Chambers. United Nations University Press.

Mumme, S., and P. Duncan. 1996. "The Commission on Environmental Cooperation and the U. S.-Mexico border environment." *Journal of Environment and Development* 5: 197–215.

Munton, D., and J. Kirton. 1994. "Environment cooperation: Bilateral, trilateral, multilateral." *North American Outlook* 4: 59–87.

Murphy, D. F., and J. Bendell. 1997. *In the Company of Partners: Business, Environmental Groups and Sustainable Development Post-Rio*. Policy Press.

Murphy, R. 1994. *Rationality and Nature: A Sociological Inquiry into a Changing Relationship*. Westview.

Neale, A. 1997. "Organizing environmental self-regulation: Liberal governmentability and the pursuit of ecological modernisation in Europe." *Environmental Politics* 6, no. 4: 1–24.

Nederveen Pieterse, J. 1995. "Globalisation as hybridisation." In *Global Modernities*, ed. M. Featherstone et al. Sage.

Nederveen Pieterse, J. 1997. "Going global: Futures of capitalism." *Development and Change* 28: 367–382.

Nelson, R. R. 1993. "A retrospective." In *National Innovation Systems*, ed. R. Nelson and N. Rosenberg. Oxford University Press.

Nelson, R. R., and N. Rosenberg, eds. 1993. *National Innovation Systems: A Comparative Analysis*. Oxford University Press.

Nguyen Cong Thanh. 1993. Viet Nam Environment Sector Study. Prepared for Asian Development Bank, Bangkok.

Nguyen Phuc Quoc. 1999. Industrial Solid Waste Prevention and Reduction: Circular Metabolism in Rubber Products and Footwear Manufacturing Industries. M.Sc. thesis, Wageningen University.

Nguyen Thi Sinh. 1998. Tinh Kha Thi cua Viec Ap Dung He Thong Quan Ly moi Truong Theo Tieu Chuan ISO 14000 doi voi Doanh Nghiep Viet Nam (The Feasibility of the Application of ISO 14000 to Viet Nam's Enterprises). B.Sc. thesis, Viet Nam National University, Ho Chi Minh City.

Nieuwenhuis, H. 1997. "De Curaçaose energiesector en de lange weg naar duurzaamheid." In *Tussen Zandstrand en Asfaltmeer*, ed. A. Mol and B. van Vliet. Jan van Arkel.

Nownes, A. J. 1991. "Interest groups and the regulation of pesticides: Congress, coalitions, and closure." *Policy Sciences* 24: 1–18.

O'Connor, D. 1994. Managing the Environment with Rapid Industrialisation: Lessons from the East Asian Experience. OECD.

O'Connor, J. 1998. *Natural Causes: Essays in Ecological Marxism.* Guilford.

OECD (Organization for Economic Cooperation and Development). 1995. OECD Environmental Performance Reviews: Japan.

OECD. 1997a. Economic Globalisation and the Environment.

OECD. 1997b. Foreign Direct Investment and the Environment: An Overview of the Literature.

OECD. 1998. Financial Market Trends no. 70.

Ohmae, K. 1995. *The End of the Nation-State.* Free Press.

Ombura, C. O. 1996. Towards an Environmental Planning Approach in an Urban Industrial Siting and Operations in Kenya: The case of Eldoret Town. Ph.D. thesis, University of Amsterdam.

Onyango, M. O. 1997. Municipal Solid Waste Management in Mombassa, Kenya. M.Sc. thesis, Centre for the Urban Environment, Rotterdam/Wageningen.

Opschoor, J. B. 1992. "Sustainable development, the economic process and economic analysis." In *Environment, Economy and Sustainable Development,* ed. J. Opschoor. Wolters-Noordhoff.

Opschoor, J. B. 1999. "Multilateral agreements on investment and the environment." Paper prepared for OECD Conference on Foreign Direct Investment and Environment, The Hague.

O'Rourke, D. J. 2000. Community Driven Regulation: The Political Economy of Pollution in Vietnam. Dissertation, University of California, Berkeley.

Paehlke, R. C. 1989. *Environmentalism and the Future of Progressive Politics.* Yale University Press.

Patel, P., and K. Pavitt. 1991. "Large firms in the production of the world's technology: an important case on 'non-globalisation.'" *Journal of international Business Studies* 1: 1–21.

Patel, P., and K. Pavitt. 1994. "The nature and economic importance of the national innovation systems." *STI Review* (OECD), no. 14: 9–31.

Paterson, M. 1996. *Global Warming and Global Politics.* Routledge.

Pearson, C. S., ed. 1987. *Multinational Corporations, Environment and the Third World: Business Matters.* Duke University Press.

Pellow, D. N., A. S. Weinberg, and A. Schnaiberg. 2000. "Putting ecological modernization to the test: Accounting for recycling's promises and performance." *Environmental Politics* 9, no. 1: 109–137.

Pepper, D. 1984. *The Roots of Modern Environmentalism.* Croom Helm.

Pepper, D. 1999. "Ecological modernisation or the 'ideal model' of sustainable development? Questions prompted at Europe's periphery." *Environmental Politics* 8, no. 4: 1–34.

Percival, D. 1996. "Country report Kenya: Beating poverty with growth." *Courier* (UNESCO), May-June 1996.

Petkova, E., and P. Veit. 2000. Environmental Accountability Beyond the Nation-State: The Implications of the Arhus Convention. World Resources Institute.

Plummer, M. G. 1997. "Regional economic integration and dynamic policy reform: The 'special' case of developing Asia." *Asia-Pacific Development Journal* 4, no. 1: 1–26.

Porter, G., and J. W. Brown. 1991. *Global Environmental Politics*. Westview.

Porter, M. E. 1990. *The Competitive Advantage of Nations*. Macmillan.

Porter, M. E., and C. van der Linde. 1995a. "Toward a new conception of the environment-competitiveness relationship." *Journal of Economic Perspectives* 9, no. 4: 97–118.

Porter, M. E., and C. van der Linde. 1995b. "Green and competitive: Ending the stalemate." *Harvard Business Review* 73, no. 4: 120–134.

Qing, D., and E. B. Vermeer. 1999. "Do good work, but do not offend the 'old communists': Recent activities of china's non-governmental environmental protection organizations and individuals." In *China's Economic Security*, ed. W. Draguhn and R. Ash. Curzon.

Ramstetter, E. 1998. "Measuring the size of foreign multinationals in the Asia-Pacific." In *Economic Dynamism in the Asia Pacific*, ed. G. Thompson. Routledge.

Rawcliffe, P. 1998. *Environmental Pressure Groups in Transition*. Manchester University Press.

Redclift, M. 1996. *Wasted: Counting the Costs of Global Consumption*. Earthscan.

Redclift, M., and T. Benton, eds. 1994. *Social Theory and the Global Environment*. Routledge.

Redclift, M., and C. Sage. 1998. "Global environmental change and global inequality: North/South perspectives." *International Sociology* 13, no. 4: 499–516.

Reed, D., ed. 1997. *Structural Adjustment, the Environment and Sustainable Development*. Earthscan.

Reinhardt, F. 1999. "Market failure and the environmental policies of firms: Economic rationales for '"beyond compliance' behavior." *Journal of Industrial Ecology* 3, no. 1: 9–21.

Reno, W. 1995. *Corruption and State Politics in Sierra Leone*. Cambridge University Press.

Richardson, J., ed. 1982. *Policy Styles in Western Europe*. Allen and Unwin.

Rinckevicius, L. 1998. Ecological Modernization and Its Perspectives in Lithuania: Attitudes, Expectations, Actions. Dissertation, Kaunas University of Technology.

Rinckevicius, L. 2000. "The ideology of ecological modernization in 'double-risk' societies: A case study of Lithuanian environmental policy." In *Environment and Global Modernity*, ed. G. Spaargaren et al. Sage.

Ringquist, E. J. 1997. "Environmental justice: Normative concerns and empirical evidence." In *Environmental Policy in the 1990s*, ed. N. Vig and M. Kraft. Congressional Quarterly Press.

RMNO (Raad voor Milieu- en Natuuronderzoek). 1996. Ruimte voor ecologische modernisering.

Roberts, J. T. 1998a. "Emerging global environmental standards: Prospects and perils." *Journal of Developing Societies* 14, no. 1: 144–165.

Roberts, J. T. 1998b. "The end of 'pollution haven' as 'comparative advantage'? Emerging international environmental standards and the Brazilian chemical industry." Paper presented at Space, Place and Nation conference, University of Massachusetts, Amherst.

Roberts, J. T., and P. E. Grimes. 1997. "World-system theory and the environment: Towards a new synthesis." Paper presented at ISA conference on Sociological Theory and the Environment, Woudschoten, Netherlands.

Robertson, R. 1992. *Globalization: Social Theory and Global Culture*. Sage.

Robertson, R. 1995. "Globalization: Time-space and homogeneity-heterogeneity." In *Global Modernities*, ed. M. Featherstone et al. Sage.

Robins, N., and S. Roberts. 1997. "Reshaping the benefits: Trade opportunities for developing-country producers from sustainable consumption and production." *Greener Management International* 19: 53–66.

Rock, M. T. 1996a. "Toward more sustainable development: The environment and industrial policy in Taiwan." *Development Policy Review* 14: 255–272.

Rock, M. T. 1996b. "Pollution intensity of GDP and trade policy: Can the World Bank be wrong?" *World Development* 24, no. 3: 471–479.

Rock, M. T. 2000. "Getting into the environment game: Ecological modernization in China and Taiwan." Paper prepared for Globalization Governance and the Environment workshop, Institute of International Studies, University of California, Berkeley.

Rowell, A. 1996. *Green Backlash: Global Subversion of the Environmental Movement*. Routledge.

Rowlands, I. H. 1995. *The Politics of Atmospheric Change*. Manchester University Press.

Rucht, D. 1993. "'Think globally, act locally'? Needs, forms and problems of cross-national cooperation among environmental groups." In *European Integration and Environmental Policy*, ed. J. Liefferink et al. Belhaven.

Rugman, A. M. 1994. "A Canadian perspective on NAFTA." *International Executive* 36, no. 1: 33–54.

Ruigrok, W., and R. van Tulder. 1995. *The Logic of International Restructuring*. Routledge.

Ruiter, W. de. 1988. "Het postmoderne Tijdperk." *Wetenschap en samenleving* 40, no. 4: 3–11.

Runge, C. F. 1994. *Freer Trade, Protected Environment: Balancing Trade Liberalization and Environmental Interests*. Council on Foreign Relations Press.

Sachs, W. 1993. "Global ecology and the shadow of 'development.'" In *Global Ecology*, ed. W. Sachs. Zed.

Sachs, W., et al. 1998. *Greening the North: A Post-Industrial Blueprint for Ecology and Equity*. Zed.

Sampson, G. P., and W. B. Chambers, eds. 1999. *Trade, Environment, and the Millennium*. United Nations Press.

Sanchez, R. A. 2000. "Governance, trade, and the environment in the context of NAFTA." Paper prepared for Globalization Governance and the Environment workshop, Institute of International Studies, University of California, Berkeley.

Sairinen, R. 2000. Regulatory Reform of Finish Environmental Policy. Dissertation, University of Technology, Helsinki.

Sarkar, S. 1990. "Accommodating industrialism: A Third World view of the West German ecological movement." *Ecologist* 20, no. 4: 147–152.

Sassen, S. 1994. *Cities in a World Economy*. Pine Forge Press.

Schmidt-Bleek, F. 1994. Wieviel Umwelt braucht der Mensch? MIPS—Das Maß für ökologisches Wirtschaften, Berlin.

Schnaiberg, A. 1980. *The Environment: From Surplus to Scarcity*. Oxford University Press.

Schnaiberg, A., A. Weinberg, and D. Pellow. 2001. "The treadmill of production and the environmental state." *Organization and Environment* 14, no. 3.

Scholte, J. A. 1996. "Beyond the buzzword: Towards a critical theory of globalisation." In *Globalisation*, ed. E. Kofman and G. Youngs. Pinter.

Scholten, L. J. R. 1998. "Stokt de Kapitaalstroom?" *Economisch-Statische Berichten*, 12 June 1998.

Schot, J. 1998. "Constructive technology assessment comes of age." In *Technology Meets the Public*, ed. A. Jamison. Aalborg University Press.

Schrijver, N. 1995. Sovereignty over Natural Resources: Balancing Rights and Duties in an Interdependent World. Ph.D. thesis, University of Groningen.

Shiva, V. 1989. *Staying Alive: Women, Ecology and Development*. Zed.

Shiva, V. 1993. "The greening of the global reach." In *Global Ecology: A New Arena of Political Conflict*, ed. W. Sachs. Zed.

Sikor, T. O., and D. O'Rourke. 1996. "Economic and environmental dynamics of reform in Vietnam." *Asian Survey* 36, no. 6: 601–617.

Situma, F. D. P. 1992. "The environmental problems in the city of Nairobi, Kenya." *African Urban Quarterly* 7, no. 1/2: 167–175.

Sklair, L. 1991. *Sociology of the Global System*. Harvester/Johns Hopkins University Press.

Sklair, L. 1994. "Global sociology and global environmental change." In *Social Theory and the Global Environment*, ed. M. Redclift and T. Benton. Routledge.

Smith, S. 1993. "The environment on the periphery of international relations: An explanation." *Environmental Politics* 2, no. 4: 28–45.

Somson, H. 1996. "The European Union and the Organization for Economic Cooperation and Development." In *Greening International Institutions*, ed. J. Werksman. Earthscan.

Sonnenfeld, D. 1996. Greening the tiger? Social Movements' Influence on the Adoption of Environmental technologies in the Pulp and Paper Industries of Australia, Indonesia and Thailand. Ph.D. thesis, University of California, Santa Cruz.

Sonnenfeld, D. 1998. "Logging versus recycling: Problems of the industrial ecology of pulp manufacturing in South-East Asia." *Greener Environment International* 22: 108–122.

Sonnenfeld, D. A. 1999. "Vikings and Tigers: Finland, Sweden and adoption of environmental technologies in Southeast Asia's pulp and paper industries." *Journal of World-Systems Research* 5, no. 1: 26–47.

Sonnenfeld, D. 2000. "Contradictions of ecological modernization: Pulp and paper manufacturing in Southeast Asia." *Environmental Politics* 9, no. 1: 235–256.

Spaargaren, G. 1997. The Ecological Modernisation of Production and Consumption: Essays in Environmental Sociology. Dissertation, Wageningen Agricultural University.

Spaargaren, G., and A. P. J. Mol. 1992. "Sociology, environment and modernity: Ecological modernisation as a theory of social change." *Society and Natural Resources* 5: 323–344.

Spaargaren, G., and A. P. J. Mol. 1995. Review of Redclift and Benton, *Social Theory and the Global Environment*. *Society and Natural Resources* 8, no. 6: 578–581.

Spaargaren, G., and B. van Vliet. 2000. "Lifestyles, consumption and the environment: The ecological modernisation of domestic consumption." *Environmental Politics* 9, no. 1: 50–76.

Spaargaren, G., A. P. J. Mol, and F. Buttel, eds. 2000. *Environment and Global Modernity*. Sage.

Spangenberg, J. H. 1995. *Towards Sustainable Europe*. Wuppertal Institute.

Spybey, T. 1996. *Globalization and World Society*. Polity Press.

Stafford, E. R., and C. L. Hartman. 1998. "Toward an understanding of the antecedents of environmentalist-business cooperative relations." In Proceedings of the American Marketing Association's Summer Educators Conference, Chicago.

Stanners, D., and P. Bourdeau, eds. 1995. Europe's Environment: The Dobris Assessment. European Environment Agency.

Staute, J. 1997. *Das Ende der Unternehmenskultur: Firmenalltag im Turbokapitalismus*. Campus.

Steenbergen, B. van. 1994. "Toward a global ecological citizen." In *The Condition of Citizenship*, ed. B. van Steenbergen. Sage.

Stern, D. I., M. S. Common, and E. B. Barbier. 1996. "Economic growth and environmental degradation: The environmental Kuznets curve and sustainable development." *World Development* 24, no. 7: 1151–1160.

Strassoldo, R. 1992. "Globalism and localism: Theoretical reflections and some evidence." In *Globalisation and Territorial Identities*, ed. Z. Mlinar. Avebury.

Streeten, P. 1993. "The special problems of small countries." *World Development* 21, no. 2: 197–202.

Switzer, J. V. 1997. *Green Backlash: The History and Politics of Environmental Opposition in the U.S.* Lynne Rienner.

Szasz, A., and M. Meuser. 1997. "Environmental inequalities: Literature review and proposals for new directions in research and theory." *Current Sociology* 45, no. 3: 99–120.

Thomas, G. M., J. W. Meyer, F. O. Ramirez, and J. Boli. 1987. *Institutional Structure: Constituting State, Society , and the Individual*. Sage.

Tobey, J. 1990. "The effects of domestic environmental policies on patterns of world trade: An empirical test." *Kyklos* 43, no. 2: 191–209.

Tomlinson. 1999. *Globalization and Culture*. Polity Press.

Trainer, T. 1988. *Developed to Death: Rethinking Third World Development*. Marshall Pickering.

Ullrich, O. 1979. *Weltniveau: In der Sachgasse der Industriegesellschaft*. Rotbuch.

UNCTAD (United Nations Conference on Trade and Development). 1988. Specific Problems of Island Developing Countries (LDC/Misc/17).

UNCTAD. 2000. World Investment Report 2000: Cross-border Mergers and Acquisitions and Development.

Underdahl, A., and O. Young. 1996. "Institutional dimensions of global change: A preliminary scoping report." Paper presented at Implementation of Environmental Treaties conference, Noordwijk, Netherlands.

Underdahl, A., M. Hisschemöller, and K. von Moltke. 1998. "The study of regime effectiveness." Agenda-setting paper, Concerted Action workshop, Noordwijk, Netherlands.

UNDP (United Nations Development Program). 1995. Incorporating Environmental Considerations into Investment Decision-making in Vietnam. UNDP, Hanoi.

UNEP (United Nations Environment Program). 1994. Background paper for Small Island Development States Conference, Barbados.

UNIDO (United Nations Industrial Development Organization). 2000. web-site http://www. unido. org/doc.

van Vliet, B. 1997. "Op weg naar een schoon eeuwfeest? De Isla-raffinaderij en het Curaçaos milieu." In *Tussen Zandstrand en Asfaltmeer*, ed. A. Mol and B. van Vliet. Jan van Arkel.

Veeken, J. M. M. 1982. Relocatie van milieuvervuilende activiteiten naar perifere landen. M.Sc. thesis, Sociaal Geografisch Instituut, Amsterdam.

Vermeer, E. B. 1998. "Industrial pollution in China and remedial policies." *China Quarterly* 156: 952–985.

Villamil, J., et al. 1971. "Open system planning: Preliminary analysis." *Northeast Regional Science Review* 1: 45–72.

Vogel, D. 1986. *National Styles of Regulation*. Cornell University Press.

Vogel, D. 1997. "International trade and environmental regulation." In *Environmental Policy in the 1990s*, ed. N. Vig and M. Kraft. Congressional Quarterly Press.

von Weizsacker, E. U., A. B. Lovins, and L. H. Lovins. 1997. *Factor Four: Doubling Wealth—Halving Resource Use*. Earthscan.

Waarden, F. van. 1995. "Persistence of national policy styles: A study of their institutional foundation." In *Convergence or Diversity*, ed. B. Unger and F. van Waarden. Avebury.

Wackernagel, M., and W. Rees. 1996. *Our Ecological Footprint: Reducing Human Impact on the Earth*. New Society Publishers.

Wallace, D. 1996. *Sustainable Industrialisation*. Royal Institute of International Affairs/Earthscan.

Wallach, L., and M. Sforza. 1999. *The WTO: Five Years of Reasons to Resist Corporate Globalization*. Seven Stories Press.

Wallerstein I. 1974. *The Modern World-System: Capitalist Agriculture and the Origins of the European World-Economy in the Sixteenth Century*. Academic Press

Wallerstein, I. 1980. *The Modern World System II: Mercantilism and the Consolidation of the European World-Economy, 1600–1750*. Academic Press

Wallerstein, I. 1990. "Culture as the ideological battleground of the modern world-system." *Theory, Culture and Society* 7: 31–55.

Wang, J. 1999. "China's national cleaner production strategy." *Environmental Impact Assessment Review* 19, no. 5/6: 437–456.

Wang, X. 2000. Information Strategy in China's Environmental Management. M.Sc. thesis, Wageningen University.

Wanyonyi, J. E. W. 1996. Towards Community and Private (Informal Sector Resource Recovery) Involvement in Solid Waste Management for Sustainable Development in Mombasa, Kenya. M.Sc. thesis, Centre for the Urban Environment, Wageningen/Rotterdam.

Wapner, P. 1995. "The state and environmental challenges: A critical exploration of alternatives to the state-system." *Environmental Politics* 4, no. 1: 44–69.

Wapner, P. 1996. *Environmental Activism and World Civic Politics*. State University of New York Press.

Wapner, P. 1997. "Governance in global civil society." In *Global Governance*, ed. O. Young. MIT Press.

Wasonga, R. G. 1999. Constructing Public-Private Correlation within a Tangible Wastewater Management Scheme: A Case Study of Kisumu Municipality, Kenya. Centre for the Urban Environment, Wageningen/Rotterdam.

Waterman, P. 1996. "Beyond globalism and developmentalism: Other voices in world politics." *Development and Change* 27: 165–18.

Waters, M. 1995. *Globalization*. Routledge.

WCED (World Commission on Environment and Development). 1987. *Our Common Future*. Oxford University Press.

Weale, A. 1992. *The New Politics of Pollution*: Manchester University Press.

Weinberg, A. 1998. "Distinguishing among green business: Growth, green, and anomie." *Society and Natural Resources* 11: 241–250.

Whatmore, S. 1994. "Global agro-food complexes and the refashioning of rural Europe." In *Globalisation, Institutions and Regional Development in Europe*, ed. A. Amin and N. Thrift. Oxford University Press.

White, G., and R. Wade. 1984. Development states in East Asia. IDS Research Report 16, Gatsby Charitable Foundation.

Wilenius, M. 1999. "Sociology, modernity and the globalization of environmental change." *International Sociology* 14, no. 1: 33–57.

Williams, M.,1995. "Rethinking sovereignty." In *Globalisation*, ed. E. Kofman and G. Youngs. Pinter.

Williams, M., and L. Ford. 1999. "The World Trade Organisation, social movements and global environmental management." *Environmental Politics* 8, no. 1: 268–289.

Windt, H. van der. 1995. *En dan: wat is natuur nog in dit land? Natuurbescherming in Nederland 1880–1990*. Boom.

World Bank. 1992. *World Development Report 1992: Development and the Environment*. Oxford University Press.

World Bank. 1993. "International trade and the environment." *World Bank Policy Research Bulletin* 4, no. 1: 1–6.

World Bank. 1994. *Adjustment in Africa: Reforms, Results and the Road Ahead*. Oxford University Press.

World Bank. 1995. Vietnam: Environmental Program and Policy Priorities for a Socialist Economy in Transition. Volume 1, report 13200-VN, World Bank.

World Bank. 1997a. *World Development Report 1997*. Oxford University Press.

World Bank. 1997b. Clear Water, Blue Skies: China's Environment in the New Century. World Bank.

WTO (World Trade Organization). 2000. WTO Annual Report 1999.

WTO-CTE (World Trade Organization Committee on Trade and Environment). 1996. "Report of the Committee on Trade and Environment." In Report of the General Council to the 1996 Ministerial Conference. WTO.

Yearley, S. 1996. *Sociology, Environmentalism, Globalisation*. Sage.

Young, O. 1989. *International Cooperation: Building Regimes for Natural Resources and the Environment*. Cornell University Press.

Young, O. 1994. *International Governance: Protecting the Environment in a Stateless Society*. Cornell University Press.

Young, O. R., ed. 1997a. *Global Governance: Drawing Insights from the Environmental Experience*. MIT Press.

Young, O. R. 1997b. "Global governance: Toward a theory of decentralized world order." In *Global Governance*, ed. O. Young. MIT Press.

Young, O. R., ed. 1999. *The Effectiveness of International Environmental Regimes: Causal Connections and Behavioral Mechanisms*. MIT Press.

Zamparutti, A., and J. Klavens. 1993. "Environment and foreign investment in Central and Eastern Europe: Results from a survey of Western corporations." In Environmental Policies and Industrial Competitiveness. OECD.

Zarsky, L. 1997. "Stuck in the mud: Nation-states, globalisation and the environment." In Globalisation and the Environment: Preliminary Perspectives. OECD.

Zarsky, L. 1999. "Havens, halos and spaghetti: Untangling the evidence about foreign direct investment and the environment." Paper prepared for OECD Conference on Foreign Direct Investment and Environment, The Hague.

Index